Biobanks in Low- and Middle-Income Countries: Relevance, Setup and Management

Karine Sargsyan • Berthold Huppertz •
Svetlana Gramatiuk
Editors

Biobanks in Low- and Middle-Income Countries: Relevance, Setup and Management

Springer

Editors
Karine Sargsyan
International Biobanking and Education
Medical University of Graz
Graz, Austria

Department of Medical Genetics
Yerevan State Medical University
Yerevan, Armenia

Ministry of Health of the Republic
of Armenia
Yerevan, Armenia

Svetlana Gramatiuk
Institute of Cellular Biorehabilition
Kharkiv, Ukraine

Berthold Huppertz
Division of Cell Biology, Histology and
Embryology, Gottfried Schatz Research Center
Medical University of Graz
Graz, Austria

ISBN 978-3-030-87639-5   ISBN 978-3-030-87637-1   (eBook)
https://doi.org/10.1007/978-3-030-87637-1

This Springer imprint is published by the registered company Springer Nature Switzerland AG.
The registered company address is: Gewerbestrasse 11, 6330 Cham, Switzerland

Preface

In recent years, there has been a growing interest in translational research building bridges between basic research and clinical research, especially in routine medicine and identification of biomarkers. Translational research is well suited to speed up drug development processes, while the costs of such processes may significantly decrease with the use of biobank facilities. Any tool and development that fosters general translational research, biomarker identification, and finally drug development is important toward advancing patient care and health. In this scenario, biobanks have become major players, and hence, biobanking has become a dynamically developing discipline. In the last few years, numerous public and private institutions have started to install and pursue biobanks on a national as well as international level.

Biobanks have become the facilitators of translational research: They have adopted the role of rapid "translators" from findings in scientific laboratories to the hospital routine. This way, biobanks have developed into the basic infrastructures for high-throughput scientific investigations. At the same time, it needs to be taken into account that this translation needs an enduring preclinical and clinical research phase.

Biobanks were installed in multiple locations, and quite a large number of these local biobanks support the idea of regional, national, and international networks. Such networks can well lead to a "sample pooling" idea to set up the best possible service for research with adequate numbers of specimens, adherence to all ethical, legal, and societal issues, access to the required clinical data, and developing tools to harmonize quality and standards.

Part of this book describes the creation of a model explaining the relationship between the preclinical and clinical phase and indicating the link to biobanking. Furthermore, this book tries to show the investigation of new trends. The final goal is to put the acquired knowledge into practice, facilitating medical experts and patients to gain usable benefit from it.

Designed as a tool for building a biorepository, this book addresses the definition of biobanking and reviews the different types of biobanks that exist, including common problems that arise. In addition, we discuss important and international biobanking societies and their strategies. We investigate topics such as sample management (e.g., collection, storage, and transfer), IT systems, sustainable biobank management, risk management, preparation of market analyses, reports, and information such as the SWOT analysis (SWOT, strengths, weaknesses, opportunities, and threats), as well as public relations and ethical aspects in the field of biobanking.

Graz, Austria
Graz, Austria
Kharkiv, Ukraine

Karine Sargsyan
Berthold Huppertz
Svetlana Gramatiuk

Acknowledgments

We would like to thank all the contributors, authors, and staff of Springer Verlag. Furthermore, we would like to say a big thank you to all staff of the Ukraine Association of Biobanks who were part of the implementation process and to the Medical University of Graz, which offers the master's course "MSc Biobanking." Without this master's course and the staff of the Medical University of Graz, and the great support of everyone else involved, this book would never have come about.

Contents

Contributors

Mykola Alekseenko Institute of Cellular Biorehabilitation, Kharkiv, Ukraine

Dominique Anderson South African Medical Research Council Bioinformatics Unit, South African National Bioinformatics Institute, University of the Western Cape, Cape Town, South Africa

Hocine Bendou South African Medical Research Council Bioinformatics Unit, South African National Bioinformatics Institute, University of the Western Cape, Cape Town, South Africa

Io Hong Cheong School of Public Health, Shanghai Jiao Tong University School of Medicine, Shanghai, China

Araz Chiloyan The Ministry of Health of the Republic of Armenia, Yerevan, Armenia

Alan Christoffels South African Medical Research Council Bioinformatics Unit, South African National Bioinformatics Institute, University of the Western Cape, Cape Town, South Africa

Svetlana Gramatiuk Institute of Cellular Biorehabilitation, Kharkiv, Ukraine

Lyudmila J. Grivtsova FGBY/FSBI (Federal state budgetary institution) "National Medical Research Center of Radiology" of the Ministry of Health of the Russian Federation, Moscow, Russia

Gabriele Hartl International Biobanking and Education, Medical University of Graz, Graz, Austria

Berthold Huppertz Division of Cell Biology, Histology and Embryology, Gottfried Schatz Research Center, Medical University of Graz, Graz, Austria

Sergey A. Ivanov FGBY/FSBI (Federal state budgetary institution) "National Medical Research Center of Radiology" of the Ministry of Health of the Russian Federation, Moscow, Russia

Andrey D. Kaprin FGBY/FSBI (Federal state budgetary institution) "National Medical Research Center of Radiology" of the Ministry of Health of the Russian Federation, Moscow, Russia

Bettina Kipperer Diagnostic and Research Center for Molecular BioMedicine, Diagnostic and Research Institute of Pathology, Medical University Graz, Graz, Austria

Zisis Kozlakidis International Agency for Research on Cancer, World Health Organization, Lyon, France

Tanja Macheiner International Biobanking and Education, Medical University of Graz, Graz, Austria

Marine Melkumova Arabkir Medical Centre, Yerevan, Armenia

Christine Mitchell International Biobanking and Education, Medical University of Graz, Graz, Austria

Heimo Müller Diagnostic and Research Center for Molecular BioMedicine, Diagnostic and Research Institute of Pathology, Medical University of Graz, Graz, Austria

Armen Muradyan Yerevan Sate Medical University, Yerevan, Armenia

Lena Nanushyan The Ministry of Health of the Republic of Armenia, Yerevan, Armenia

Vladimir A. Petrov FGBY/FSBI (Federal state budgetary institution) "National Medical Research Center of Radiology" of the Ministry of Health of the Russian Federation, Moscow, Russia

Karine Sargsyan International Biobanking and Education, Medical University of Graz, Graz, Austria
Department of Medical Genetics, Yerevan State Medical University, Yerevan, Armenia
Ministry of Health of the Republic of Armenia, Yerevan, Armenia

Sergey Sargsyan Arabkir Medical Centre, Yerevan, Armenia

Tamara Sarkisian Department of Medical Genetics, Yerevan State Medical University, Yerevan, Armenia

Erik Steinfelder Thermo Fisher Scientific, Waltham, MA, USA

Werner Strasser fragmentiX Storage Solutions GmbH, Klosterneuburg, Austria

Konstantin Yenkoyan Yerevan State Medical University, Yerevan, Armenia

Bernhard Zatloukal fragmentiX Storage Solutions GmbH, Klosterneuburg, Austria

Kurt Zatloukal Diagnostic and Research Center for Molecular BioMedicine, Diagnostic and Research Institute of Pathology, Medical University Graz, Graz, Austria

Abbreviations

AMRI	Alliance of Medical Research Infrastructures
BBMRI	Biobanking and Biomolecular Resources Research Infrastructure
BBMRI-ERIC	Biobanking and Biomolecular Resources Research Infrastructure—European Research Infrastructure Consortium
BEMT	Biobank Economic Modeling Tool
BIMS	Biobank Information Management System
CEB	Commission Ethical and Bio-ethical
CFB	Consent Form Biobank
CNS	Central nervous system
CRC	Colorectal cancer
CSF	Central spinal fluid
CSS	Computationally secure secret sharing
CTRNet	Canadian Tissue Repository Network
DNA	Deoxyribonucleic acid
DTA	Data Transfer Agreement
EATRIS-ERIC	European Advanced Translational Research Infrastructure in Medicine-ERIC
EBW	European Biobank Week
EC	European Commission
ECRIN-ERIC	European Clinical Research Infrastructure Network-ERIC
EFGCP	European Forum for Good Clinical Practice
EFPIA	European Federation of Pharmaceutical Manufacturers and Associations
EGF	Ethics and Governance Framework
EGFR	Epidermal growth factor receptor
EH	Essential hypertension
ELSI	Ethical, Legal, and Societal Issues
EPIC	European Prospective Investigation into Cancer and Nutrition
ERIC	European Research Infrastructure Consortium
ESBB	European, Middle Eastern and African Society for Biobanking
ESFRI	European Strategy Forum on Research Infrastructures
FFPE	Formalin-fixed, paraffin-embedded tissue
FMF	Familial Mediterranean fever

FP7	Seventh Framework Programme
GDPR	General Data Protection Regulation
GLP	Good laboratory practice
GWA	Genome-wide association
G2P	Genotype-to-phenotype
HTA	Human Tissue Act
HR	Human resources
H3 Africa	Human Heredity and Health in Africa Consortium
IARC	International Agency for Research on Cancer
IATA	International Air Transport Association
IBBL	Integrated Biobank of Luxembourg
IGF-1	Insulin-like growth factor-1
IPR	Intellectual Property Rights
IRB	Institutional Review Board
ISBER	International Society for Biological and Environmental Repositories
IST	Information-Theoretic Security
ISO	International Organization for Standardization
JRC	Joint Research
KPIs	Key performance indicators
LIMS	Laboratory Information Management System
LMICs	Low- and middle-income countries
MSCs	Mesenchymal stromal/stem cells
MTA	Material Transfer Agreement
NASBIO	National Association of Biobanks and Biobanking Specialists
NCI	National Cancer Institute
NCDs	Non-communicable diseases
NGS	Next-generation sequencing
OBBR	Office of Biobank and Biological Specimen Research
OECD	Organisation for Economic Co-operation and Development
PC	Prostate cancer
PCR	Polymerase chain reaction
PCSM	Prostate cancer-specific mortality
PDC	Patient-derived cell
PDX	Patient-derived xenograft
PIL	Patients Information Letter
PR	Public relations
PSS	Perfect Secret Sharing
P3G	Public Population Project
QMS	Quality management system
REC	Regional Ethics Committee
RI	Research institution
RIA	Research and innovation actions
RNA	Ribonucleic acid
RP	Retinitis pigmentosa

RRI	Responsible research and innovation
RTD	Research and technological development
SAC	Sample Access Committee
SAPs	Several Sample Access Policies
SCUAB	Steering Committee of the Ukraine Association of Biobanks
SMART	Specific, Measurable, Achievable, Relevant, Time Bound
SMPC	Secure Multi-party Computation
SOP	Standard Operating Procedures
SPIDIA	Standardisation and Improvement of Generic Preanalytical Instruments and Procedures for In Vitro Diagnostics
SSS	Shamir's Secret Sharing
SWOT	Strengths, Weaknesses, Opportunities, and Threats
TOD	Target Organ Damage
TC	Technical Committee
UAB	Ukraine Association of Biobank
UNESCO	United Nations Educational, Scientific and Cultural Organization
VIL	Volunteers Information Letter
WHO	World Health Organization

Introduction to Biobanking

Karine Sargsyan and Berthold Huppertz

Abstract

Human biobanks are collections of biological specimens of a humanoid nature (physical materials of humans) that are annotated with appropriate clinical, lifestyle, laboratory, and other data from their donors. The link between bio-specimens and data makes specimen collections in biobanks significant. A mere collection of human biological material without corresponding data is not a biobank. Once a biobank is able to confirm and provide appropriate medical and other data from a donor and link it to a bio-specimen, it can be called a biobank. Around the world, biobanks collect, systematize, and store biological samples in direct relation to their (clinical) data, both of which are of key scientific importance for medical and pharmacological research. Most biobanks focus on routine research infrastructure tasks, collecting tissues and biological samples from primary sources, and sharing them with research groups.

Keywords

Human bio-samples · Biological material · Biorepository · Data collection · Types of biobanks · Definition of biobanks

K. Sargsyan
International Biobanking and Education, Medical University of Graz, Graz, Austria

Department of Medical Genetics, Yerevan State Medical University, Yerevan, Armenia

Ministry of Health of the Republic of Armenia, Yerevan, Armenia

B. Huppertz (✉)
Division of Cell Biology, Histology and Embryology, Gottfried Schatz Research Center, Medical University of Graz, Graz, Austria
e-mail: berthold.huppertz@medunigraz.at

The progress of medical and pharmaceutical research in the world directly depends on the quality of human bio-samples. At the same time, the slowdown in the pace of development during the last two decades could directly be linked to the limited availability of high-quality samples. Only the material from a large number of patients, collected by structured biobanks, has managed to make analysis of the prevalence of diseases, the severity of syndromes and symptoms, as well as the molecular and genetic patterns possible.

Biobanks around the world are collecting, systematizing, and storing biological samples directly linked to their (clinical) data, both of which are of pivotal scientific importance for medical and pharmacological research. Materials stored in biobanks are very multifaceted: from tissues including normal and pathological tissues to all kinds of body fluids including blood, serum, and urine, to isolated molecules such as deoxyribonucleic acid (DNA) and ribonucleic acid (RNA), and to primary cells and cell lines [1, 2].

Biobanks differ in structure and in the specialization of the biological material to be stored and made available: Some biobanks collect hard-to-reach tissue types such as eyes, brain, and bones, while others emphasize on revealing and separating cell lines from donor blood or tissues [3]. Most of the biobanks focus on the routine task of the research infrastructure and collect tissues and biological samples from primary sources and distribute them to research groups [2]. Biobanks play a prominent role in research efforts in a variety of fields including oncology, virology, and medical genetics. In these fields, studies aim at identifying key mechanisms for the development and progression of diseases, signalling molecules, key proteins, and genes involved in the evolution of diseases. The implementation of this innovative knowledge to identify (mostly in the early stages of disease) and disclose treatments for diseases is in a straightforward relationship with biobanks. For instance, upstream pharmaceutical companies use fluorescence-based visualization techniques to identify specific DNA and RNA sequences as well as additional methods including organ-on-a-chip approaches, single-cell RNA sequencing, and genetic screening (next-generation sequencing, NGS) in an attempt to identify new and promising biomarkers associated with specific diseases [4].

The lack of sustainability including the lack of long-term financing (public or private) for the development and support of biobanks and the high demand of bio-specimens are key factors that cause the slow growth in the field of biomedical research. This is especially true for the increasing amount of high-quality bio-specimens and related data needed in biomedical research studies. Biobanks experience a profusion of economic issues affecting the capabilities and capacities to offer bio-specimens and services sustainably [4–7].

Moreover, numerous biorepositories have difficulties in calculating the real overall charge rate of providing analysis services and costs of allocating samples and data. Some also have difficulties in determining a fair cost remuneration (e.g., fair consumer costs) [8], all of which lines up the market and produces a factual rate of biobanking to store single bio-specimen and data. This resulted in the development and implementation of biobank governing and managing models mostly based on cost recovery. However, such calculations are still among the major operative problems of several biobanks all over the world [7, 9–15].

The perception of the possibility to earn fast funds and get financial growth by implementing and building biobanks is exacting and does not fittingly reflect general and particular ideas of biobanks, especially in developing countries.

The novelty of this book is the consideration that examines the evolution of a biobanking economic plan in the Ukraine, with possible beneficial considerations for feasible subsidy tools (e.g., standards for the repayment of expenses) to insure biobanks' financial and economic sustainability [3].

The aims of the Ukrainian biobanking project are the assignment and the accomplishment of a compressed sustainability plan containing:

- Immediate approach to address proximate monetary issues
- Long-standing monetary and functional sustainability done over centralized operation growth and streamlined trading operations

Definitions of a Biobank

Human biobanks are elucidated as collections of biological specimens of humanoid nature (physical materials of human being) that are accompanied with corresponding clinical, lifestyle, laboratory, and other data of their donors. Examples of corporeally substances (e.g., biological samples) are blood, cells, tissues, and typical genetic materials (DNA, RNA). Reliant on the intention of certain biorepositories, equally genetic data and other relevant clinical data and lifestyle-narrated data of the donor may be accompanied to the collected and stored specimen. The connection of bio-specimen and data makes the specimen collections in biobanks significant [16–19].

Just a collection of human biological material without corresponding data is not a biobank. If a biobank can confirm and provide corresponding medical and other data from a donor and link it to a bio-specimen, then it can be called a biobank. However, in addition this institution still needs to follow workplace excellence regulations from blood standards to the standard operating procedures (SOP) of proceeding genetic analyses. The data collection of these well-functioning biobanks can be made available to researchers and/or research groups whenever a new study on human biological material is started. Here, all necessary ethical concerns need to be taken into account, and written informed consents of the donors need to be in place whenever needed. Also, advanced databases need to be run, and applications for samples and data always need to be embedded in a framework of data security and data safety, especially if genetic data are used [18, 20].

Biorepositories are adopting applications for diagnostic measurements or curative intentions. These types of biobanks are described in this book only if they include a complete or partial application in biomedical research. Aimed at the given consideration, blood banks, which may also contain umbilical cord blood, and usual blood banks, along with organ banks collecting samples, tissues, and organs for medical applications, are outside of the scope of this book. This is similarly true for biobanks installed for argumentative intention (withstand iniquity—forensic), or for nonphysical uses of diagnostic procedures in genetics, for example, for the association to test fatherhood. A conclusive combination of exclusion criteria for biobanks to be discussed in this book has also revealed that biobanks made only for the plan (as in the expansion of population genetics or developmental science) are not concerned with physical examinations. However, the discovery of the arrangement of dissimilarities in genetics in people and the discovery of the phylogeny of the human will also not be considered [10, 12, 16, 21–24].

The consideration for a restriction of the groups of biobanks included in this book is based on a specific focus of distinctive and general concern, which emphasizes biorepositories intended for biomedical research studies in the Ukraine, describing the development of those biobanks that are in focus of this book. The challenge, however, is to manage this interest so that it can be implemented with certainty. Donors of biological specimens require, to a higher degree than in other areas, acknowledged permission not only to residual biological samples but also to the extensive medical and corresponding data. The considerations indispensably necessitate, by this position, to explain the special treatment and principle of biobanks for biomedical studies [8, 25].

Human biobanks are important innovative basic infrastructures for medical research, as they can act as essential partners in a distinctive disease science not merely for single patients but on an epidemiological side too, together with the elaboration of diagnoses, prophylactics, and therapy claims and methods. In theory, human biobanks are able to act as the main basis of a precious contribution to fight large and commonly known diseases. Biobanks can also aggregate a considerable amount of information in one institution to reveal specific particularities of a pharmacological agent (patients) or for certain illnesses (pharmacogenetics and pharmacogenomics) and still be important after several decades of collection [16, 17, 26, 27].

Biorepositories, which assist biomedical sciences, can put huge efforts in largeness and construction. Biobanks show particularly valuable effects in biomedical sciences and hence are considered to be of general interest in developed countries. In these biobanks, specimens and data from a comprehensive part of the population are usually stored long-term, which is mostly accepted by the general population. Such structured collections are easily qualified for application in biomedical research. However, there is still skepticism whether biobanks can collect specimens and data without a specific collection strategy based on a specific scientific goal. The skeptics are worried about an "indefinite" collection at a biobank and about the danger to include such a comprehensive plan in a defective biobanking apparatus for future studies [16].

Biobanks that are used for biomedical research benefit not only from knowledge discovery but often also have a distinctive effect on the contributors that supply specimens and corresponding data. Donors who contribute to an existing clinical trial by supplying medically applicable human biological material (e.g., tumor pieces) are confident that they and others suffering from the same disease will benefit from the practical advancement. Such a strategy will make treatments available to the patients that may be very expensive or simply not available for the usual treatment if collections in biobanks would not be available [16].

References

1. Barnes, R. O., Parisien, M., Murphy, L. C., & Watson, P. H. (2008). Influence of evolution in tumor biobanking on the interpretation of translational research. *American Association for Cancer Research, 17*(12), 3344–3350. https://doi.org/10.1158/1055-9965
2. Organization for Economic Cooperation and Development. (2007). *OECD best practices guidelines for biological resource centers.* Retrieved from http://www.oecd.org/sti/emerging-tech/oecdbestpracticeguidelinesforbiologicalresourcecentres.htm
3. Sudlow, C., Gallacher, J., Allen, N., Beral, V., Burton, P., Danesh, J., Downey, P., Elliott, P., Green, J., Landray, M., Liu, B., Matthews, P., Ong, G., Pell, J., Silman, A., Young, A., Sprosen, T., Pearkman, T., & Collins, R. (2015). UK biobank: An open access resource for identifying the causes of a wide range of complex diseases of middle and old age. *PLoS Medicine, 12*(3). https://doi.org/10.1371/journal.pmed.1001779
4. Filbin, M. G., & Mario, L. S. (2016). Gliomas genomics and epigenomics: Arriving at the start and knowing it for the first time. *Annual Review of Pathology Mechanisms of Disease, 11,* 497–521. https://doi.org/10.1146/annurev-pathol-012615-044208
5. Mackenzie, J. (2011). The old care paradigm is dead, long live the new sustainable care paradigm: How can GP commissioning consortia meet the demand challenges of 21st century healthcare? *London Journal of Primary Care, 4,* 64–68.
6. Macheiner, T., Huppertz, B., Bayer, M., & Sargsyan, K. (2017). Challenges and driving forces for business plans in biobanking. *Biopreservation and Biobanking, 15*(2), 121–125. https://doi.org/10.1089/bio.2017.0018
7. Odeh, H., Miranda, L., Rao, A., Vaught, J., Greenman, H., McLean, J., Reed, D., Memon, S., Fombonne, B., Guan, P., & Moore, H. M. (2015). The biobank economic modeling tool (BEMT): Online financial planning to facilitate biobank sustainability. *Biopreservation and Biobanking, 16*(6), 421–429. https://doi.org/10.1089/bio.2015.0089
8. Sargsyan, K., Macheiner, T., Story, P., Strahlhofer-Augsten, M., Plattner, K., Riegler, S., Granitz, G., Bayer, M., & Huppertz, B. (2015). Sustainability in biobanking: Model of biobank Graz. *Biopreservation and Biobanking, 13*(6), 410–420. https://doi.org/10.1089/bio.2015.0087
9. Kauffmann, F., & Cambon-Thomsen, A. (2008). Tracing biological collections: Between books and clinical trials. *JAMA, 299*(19), 2316–2318. https://doi.org/10.1001/jama.299.19.2316
10. Simeon-Dubach, D., & Watson, P. (2014). Biobanking 3.0: Evidence based and customer focused biobanking. *Clinical Biochemistry, 47*(4–5), 300–308. https://doi.org/10.1016/j.clinbiochem.2013.12.018
11. Watson, P. H., Nussbeck, S. Y., Cater, C., O'Donoghue, S., Cheah, S., Matzke, L. A. M., Barnes, R. O., Bartlett, J., Carpenter, J., Grizzle, W. E., Johnston, R. N., Mes-Masson, A.-M., Murphy, L., Sexton, K., Shepherd, L., Simeon-Dubach, D., Zeps, N., & Schacter, B. (2014). A framework for biobank sustainability. *Biopreservation and Biobanking, 12*(1), 60–68. https://doi.org/10.1089/bio.2013.0064
12. Cormier, C. Y., Mohr, S. E., Zuo, D., Hu, Y., Rolfs, A., Kramer, J., Taycher, E., Kelley, F., Fiacco, M., Turnbull, G., & LaBaer, J. (2010). Protein structure initiative material repository: An open shared public resource of structural genomics plasmids for the biological community. *Nucleic Acids Research, 38,* 743–749. https://doi.org/10.1093/nar/gkp999
13. Cormier, C. Y., Park, J. G., Fiacco, M., Steel, J., Hunter, P., Kramer, J., Singla, R., & LaBaer, J. (2011). PSI: Biology-materials repository: A biologist's resource for protein expression plasmids. *Journal of Structural and Functional Genomics, 12*(2), 55–62. https://doi.org/10.1007/s10969-011-9100-8

14. Seiler, C. Y., Park, J. G., Sharma, A., Hunter, P., Surapaneni, P., Sedillo, C., Field, J., Algar, R., Price, A., Steel, J., Throop, A., Fiacco, M., & LaBaer, J. (2014). DNASU plasmid and PSI: Biology-materials repositories: Resources to accelerate biological research. *Nucleic Acids Research, 42*, D1253–D1260. https://doi.org/10.1093/nar/gkt1060
15. Lamesch, P., Li, N., Milstein, S., Fan, C., Hao, T., Szabo, G., Hu, Z., Venkatesan, K., Bethel, G., Martin, P., Rogers, J., Lawlor, S., McLaren, S., Dricot, A., Borick, H., Cusick, M. E., Vandenhaute, J., Dunham, I., Hill, D. E., & Vidal, M. (2007). hORFeome v3.1: A resource of human open reading frames representing over 10,000 human genes. *Genomics, 89*(3), 307–315. https://doi.org/10.1016/j.ygeno.2006.11.012
16. German Ethics Council. (2010). *Human biobanks for research: Opinion.* Retrieved from https://www.ethikrat.org/fileadmin/Publikationen/Stellungnahmen/englisch/DER_StnBiob_Engl_Online_mitKennwort.pdf
17. Yang, X., Boehm, J., Yang, X., Salehi-Ashtiani, K., Hao, T., Shen, Y., Lunonja, R., Thomas, S. R., Alkan, O., Bhimdi, T., Green, T. M., Johannessen, C. M., Silver, S. J., Nguyen, C., Murray, R. R., Hieronymus, H., Balcha, D., Fan, C., . . . Root, D. E. (2011). A public genome-scale lentiviral expression library of human ORFs. *Nature Methods, 8*, 659–661. https://doi.org/10.1038/nmeth.1638
18. Knoppers, B. M., Fortier, I., Legault, D., & Burton, P. (2008). Population genomics: The public population project in genomics (P^3G): A proof of concept? *European Journal of Human Genetics, 16*, 664–665. https://doi.org/10.1038/ejhg.2008.55
19. Doucet, M., Becker, K. F., Björkman, J., Bonnet, J., Clément, B., Daidone, M.-G., Duyckaerts, C., Erb, G., Haslacher, H., Hofman, P., Huppertz, B., Junot, C., Lundeberg, J., Metspalu, A., Lavitrano, M., Litton, J.-E., Moore, H. M., Morente, M., Naimi, B.-Y., . . . Dagher, G. (2016). Quality matters: 2016 annual conference of the National Infrastructures for biobanking. *Biopreservation and Biobanking, 15*(3), 270–276. https://doi.org/10.1089/bio.2016.0053
20. Schenk, M., Huppertz, B., Obermayer-Pietsch, B., Kastelic, D., Hörmann-Kröpfl, M., & Weiss, G. (2016). Biobanking of different body fluids within the frame of IVF—A standard operating procedure to improve reproductive biology research. *Journal of Assisted Reproduction and Genetics, 34*, 383–290. https://doi.org/10.1007/s10815-016-0847-5
21. Park, J., Hu, Y., Murthy, T. V. S., Vannberg, F., Shen, B., Rolfs, A., Hutti, J. E., Cantley, L. W., LaBear, J., Harlow, E., & Brizuela, L. (2005). Building a human kinase gene repository: Bioinformatics, molecular cloning, and functional validation. *PNAS, 102*(23), 8114–8119. https://doi.org/10.1073/pnas.0503141102
22. Yu, X., Bian, X., Throop, A., Song, L., Del Moral, L., Park, J., Seiler, C., Fiacco, M., Steel, J., Hunter, P., Saul, J., Wang, J., Qiu, J., Pipas, J. M., & LaBaer, J. (2014). Exploration of Panviral proteome: High-throughput cloning and functional implications in virus-host interactions. *Theranostics, 4*(8), 808–822. https://doi.org/10.7150/thno.8255
23. Brat, D. J., Verhaak, R. G., Aldape, K. H., Yung, W. K. A., Salama, S. R., Cooper, L. A. D., Rheinbay, E., Miller, C. R., Vitucci, M., Morozova, O., Robertson, A. G., Noushmehr, H., Laird, P. W., Cherniack, A. D., Akbani, R., Huse, J. T., Ciriello, G., Poisson, L. M., Barnholtz-Sloan, J. S., . . . Zhang, J. (2015). Comprehensive, integrative genomic analysis of diffuse lower-grade gliomas. *New England Journal of Medicine, 372*(26), 2481–2498. https://doi.org/10.1056/NEJMoa1402121
24. Colledge, F., Elger, B., & Howard, H. C. (2013). A review of the barriers to sharing in biobanking. *Biopreservation and Biobanking, 11*(6), 339–346. https://doi.org/10.1089/bio.2013.0039
25. Colledge, F., Persson, K., Elger, B., & Shaw, D. (2013). Sample and data sharing barriers in biobanking: Consent, committees, and compromises. *Annals of Diagnostic Pathology, 18*(2), 78–81. https://doi.org/10.1016/j.anndiagpath.2013.12.002
26. Moser, G., & Huppertz, B. (2017). Implementation and extravillous trophoblast invasion: From rare archival specimens to modern biobanking. *Placenta, 56*, 19–26. https://doi.org/10.1016/j.placenta.2017.02.007

27. Fortin, S., Pathmasiri, S., Grintuch, R., & Deschênes, M. (2011). 'Access arrangements' for biobanks: A fine line between facilitating and hindering collaboration. *Public Health Genomics, 14*(2), 104–114. https://doi.org/10.1159/000309852

Scientific Relevance, Problems and the Destitution for Principles of Biobanks and Biobanking Activities

2

Svetlana Gramatiuk and Karine Sargsyan

Abstract

There are countless collections of human biological material in the Ukraine. They are collected, stored and used for various purposes, mostly in scientific organizations without recognizable, clear principles. It is becoming increasingly apparent that in the Ukraine a cooperation between academic institutions and private pharmaceutical or biotechnology companies is possible. Furthermore, due to the extensive growth of biobanks, many investigators, universities, societies and organizations have been engaged in the creation of harmonizing and standardizing biobanks. This chapter lists the different problems of biobanks in the Ukraine and shows the need to define precise standard principles in the field of biobanking and the need for structures that can provide a safe and harmonious arrangement. In addition, there are similar but not identical definitions of biorepositories and different relevant concepts for the validation, guidance, management, implementation, authorization, application and possible cessation of human biorepositories and databases.

Keywords

Limited standard principles · Data management · Data protection · Harmonization · International societies and networks · OECD · ISBER · European Commission · ESBB · BBMRI-ERIC

S. Gramatiuk
Institute of Cellular Biorehabilitation, Kharkiv, Ukraine

K. Sargsyan (✉)
International Biobanking and Education, Medical University of Graz, Graz, Austria

Department of Medical Genetics, Yerevan State Medical University, Yerevan, Armenia

Ministry of Health of the Republic of Armenia, Yerevan, Armenia
e-mail: karine.sargsyan@medunigraz.at

9

All types of human biological material are stored in the Ukraine. They are collected, stored and used for various purposes when medical expertise is already available and ready to develop new drugs. Deficiency and/or limited standard principles in this domain have always been present in the Ukraine. In this country, countless collections are present, varying in size, intention and model, mostly located in scientific organizations with no discernible sharp principles. University departments or the pharmaceutical industries have installed collections of human biological material with different goals based on diagnosis, used therapeutics, specific studies or clinical validation. It is becoming more and more apparent that a cooperation between academic institutions and private pharmaceuticals or biotechnology companies in the Ukraine is possible and needed. In this respect, the need to define a precise principle is of utmost importance [1–5].

Following today's methods of electronic data processing in routine clinical practice, the instructions in biobanks with regard to data management and the possibilities of disseminating this data are usually expanded and adapted quickly. This expansion raises a vivid question about the adaptation of principles including data safety and data security as well as data protection of donors.

Although biobanks are preceded by important innovations in biomedical and pharmaceutical science, they may also cause unease and mistrust. In the Ukraine, there is a strong interest of patient and donor representatives who want to be part of the process of defining biobanks and not simply leave the biobankers work and decide single-handedly. Their concern is a random usage of specimens and their corresponding annotations. An additional problem is that possible contributors may become pressured to oversubscribe and "donate" samples or may even obtain physical damage from medical manipulation [2]. Although these problems are independent of whether genetic or other data is associated with the sample, the representatives believe that genetic analysis is always possible. Hence, it is their belief that biobanks are able to produce clear evidence-based data on the individuality of the donor, which can negatively affect the donor. Therefore, apart from the subject of data defined justice, the donors and their hereditary blood relatives need to be efficiently secured against hereditary severity and stigma. Biospecimen preservation from acute cases only can concern a specific population assembly, when the specimens are combined using medical data in biobanks [2, 3, 6–9].

Rarely, particular key players and suppliers take part in the debate surrounding human samples and essential data. Over the last decade, international discussions have shown that biobanks entail a variety of ethical, constitutional and societal tasks [2]. There is a clear need for structure that can provide a safe and harmonious arrangement to defend biobanks in the Ukraine against opponents who have a negative opinion, especially since the trend shows that researchers from different countries are becoming more and more comparable.

The tissue collections, with the purpose of medical diagnosis or therapeutics founded at the beginning, have now gained tremendous value for exploration, which enables the application to newer analytical methods in molecular biology. In reality, to accomplish this, it is necessary to change the regulation and enforcement of collections in the Ukraine. Donors often would not tolerate usage of their samples in studies, especially in certain genetic studies [2, 3, 8, 9]. This issue is not restricted to the Ukraine but has become apparent in most developed countries with the respective tissue collections for diagnosis rather than for scientific purposes.

For an optimal use of the network of biobanks in the Ukraine, it may be appropriate to provide medical clinical data and their applications from all the different fields. With the development of modern technologies, data collected in biorepositories can be exchanged and bundled via a network. It can lead to a size and style that goes beyond those collected in the past by local donors. Of course, rules for data safety, data security and data protection of donors as well as for hindering uncontrolled usage of samples and data, followed by the inclusion into structured biobanks, need to be considered [2].

In the structure of biobanks, where the complexity of the work on site includes in and outgoing samples and data on a daily basis, the responsibility for the community to store, handle and distribute biospecimens and corresponding annotations is a priority. Administrative disputes have to be differentiated: it is about providing a corresponding "chain of responsible persons" in order to be able to transparently show the performance and the correspondence to the regulation throughout the entire length of the system. This is especially important for donor safety [10–13].

Another question is whether and under what specific conditions the samples and data as well as the services to third parties should be provided to academic and private scientists for studies from the Ukrainian biorepositories and under which circumstances can human biological materials and corresponding data be possibly destroyed.

A final point that remains to be determined concerns the progress that can be achieved if the biobank ceases to exist. It is required to have instructions on actions based on the handling of the stored human samples and data.

Collections of samples and data may be operated based on in-sector organizations or also by persons or private institutions. Regardless of this, it needs to be defined how such collections may progress into biobanks. In addition, biobanks can be either non-profit-oriented or profitable. The separate authorized system and the announced objective may require different regulatory frameworks.

In terms of the organizationally regulated biobanks of the Ukraine, a balance must be made between all the different interests already mentioned. In medical research and on behalf of innovation, frameworks must be created that allow donors who wish to contribute to biobanks to be placed in a regulatory framework to ensure optimal use of human biological material and corresponding annotation in biomedical sciences [2].

Simultaneously, parallel to the above-mentioned measures, restrictions are needed in order to protect donors and further involved people against danger, which can be related to the instalment and operational process of biobanks [2].

In recent times, many investigators, universities, societies and organizations have been engaged in the definition of biobanks in various ways. Important international societies such as the Organisation for Economic Co-operation and Development (OECD); the International Society for Biological and Environmental Repositories (ISBER); the European, Middle Eastern and African Society for Biobanking

(ESBB); and the Biobanking and Biomolecular Resources Research Infrastructure-European Research Infrastructure Consortium (BBMRI-ERIC) have articulated their definitions.

The above-mentioned organizations offer similar but not identical definitions of biorepositories: biobanks are organized infrastructures storing a set of biologically significant samples and the combined data to create the linkage between patients and population on one side with science and drug development on the other side [6, 14].

Specialized international organizations [14–18] contribute relevant concepts for the validation, guidance, management, implementation, authorization, application and possible cessation of human biorepository and genetic study databases, which are structured substitutions that can be used for hereditary research. Such databases contain information on the stored human biological samples and/or analytical data when analysing the same samples, and will be extended with the help of an intelligent biobank network.

The international society ISBER defines a biobank as follows: "*defined as an actual or virtual entity that may receive, process, store or distribute specimens in support of a study or multiple studies and their associated data*" [19].

In 2015, the European Commission (EC) and Joint Research Centre (JRC) stated in "Biobanking in Europe" (a scientific and technical report): these are forecasts for synchronizing and establishing contacts that set business description. Biobanks or biorepositories are systematic collections that consist of human biological specimens and combined annotation, which have an inordinate meaning for science in general and, the hot topic of biomedical sciences, personalized medicine, in particular [20]. Furthermore, the European Commission issued an additional definition. The statement, which was made at an international meeting on ethics and regulation related to international research in the field of biobanks—Biobanks for Europe, included an extensive explanation for biobanks.

Biobanks:

- Collect and store human biological samples, which are linked to clinical and medical data, and frequently to epidemiologic data.
- Are usually not stable, may be non-sustainably "planned" study cohorts collecting biological materials and corresponding data on a constant or extended basis.
- Are accompanied with accurate and/or continuously developing study plans (period of sample collection).
- Have installed pseudonymization or anonymization processes to ensure patients' privacy. However, there may still be a precise mode of possibility procedure to reidentify the donor in the direction from research data back to the donor, to intelligently track back if clinically relevant data for a patient appears.
- Comprise installed organizational control mechanisms and measures (e.g. consent) that aim to save patients' privacy and rights as well as stakeholders' motivation.

The largest network of biobanks, the Pan-European Biobank network called Biobanking and Biomolecular Resources Research Infrastructure (BBMRI), explains biobanks as a medical research infrastructure: biobanks collect biological specimens and combine data, which is only a primary matter for progression and development of biotech, enabling effective biomolecular and biomedical research [20, 21].

In addition, many other definitions of biobanks exist, which are unfortunately of inconstant value—biobank definitions often focus on only one or several particular characteristics of biobanks. The much unused but widely accepted description/definition comes from Kauffmann and Cambon-Thomsen in 2008 [22]: a biobank is "*an organized collection of human biological material and associated information stored for one or more research purposes*". In conclusion, a description presented by Artene et al. relates to the biorepositories with two distinct abilities:

- Collecting, proceeding and storing material and data.
- Database is smart in terms of demographic and medical data, which are analysed individually and inserted in a register with concrete measurement units. A biological sample is therefore characterized by analytical data, phenotypic data and medical and treatment data. These data are stored in long-term storage facilities and used by researchers [23].

In general, a biobanks' definition contains three sets of comparatively different instructions:

- Human biological specimens
- Data that is loaded into the system
- Ethical and legal issues like the informed consent to assure privacy of donors

All this makes scientific biobanks an interdisciplinary system of interaction between science and practice and a universal approach in all phases of the cycle.

References

1. Lamesch, P., Li, N., Milstein, S., Fan, C., Hao, T., Szabo, G., Hu, Z., Venkatesan, K., Bethel, G., Martin, P., Rogers, J., Lawlor, S., McLaren, S., Dricot, A., Borick, H., Cusick, M. E., Vandenhaute, J., Dunham, I., Hill, D. E., & Vidal, M. (2007). hORFeome v3.1: A resource of human open reading frames representing over 10,000 human genes. *Genomics, 89*(3), 307–315. https://doi.org/10.1016/j.ygeno.2006.11.012
2. German Ethics Council. (2010). *Human biobanks for research: Opinion.* Retrieved from https://www.ethikrat.org/fileadmin/Publikationen/Stellungnahmen/englisch/DER_StnBiob_Engl_Online_mitKennwort.pdf
3. Luque, J. S., Quinn, G. P., Montel-Ishino, F. A., Arevalo, M., Bynum, S. A., Noel-Thomas, S., Wells, K. J., Gwede, C. K., Meade, C. D., & Tampa Bay Community Cancer Network Partners. (2012). Formative research on perceptions of biobanking: What community members think. *Journal of Cancer Education, 27*(1), 91–99. https://doi.org/10.1007/s13187-011-0275-2

4. Mascalzoni, D., Dove, E. S., Rubinstein, Y., Dawkins, H. J., Kole, A., McCormack, P., Woods, S., Riess, O., Schaefer, F., Lochmüller, H., Knoppers, B. M., & Hansson, M. (2016). International charter of principles for sharing biospecimens and data. *European Journal of Human Genetics, 24*(7), 1096. https://doi.org/10.1038/ejhg.2015.237

5. Gottweis, H., & Zatloukal, K. (2007). Biobank governance: Trends and perspectives. *Pathobiology, 74*(4), 206–211. https://doi.org/10.1159/000104446

6. Organization for Economic Cooperation and Development. (2007). *OECD practices guidelines for biological resource centers.* Retrieved from http://www.oecd.org/sti/emerging-tech/oecdbestpracticeguidelinesforbiologicalresourcecentres.htm

7. Knoppers, B. M., Fortier, I., Legault, D., & Burton, P. (2008). Population genomics: The public population project in genomics (P^{3}G): A proof of concept? *European Journal of Human Genetics, 16*, 664–665. https://doi.org/10.1038/ejhg.2008.55

8. Colledge, F., Elger, B., & Howard, H. C. (2013). A review of the barriers to sharing in biobanking. *Biopreservation and Biobanking, 11*(6), 339–346. https://doi.org/10.1089/bio.2013.0039

9. Colledge, F., Persson, K., Elger, B., & Shaw, D. (2013). Sample and data sharing barriers in biobanking: Consent, committees, and compromises. *Annals of Diagnostic Pathology, 18*(2), 78–81. https://doi.org/10.1016/j.anndiagpath.2013.12.002

10. Braun, L., Lesperance, M., & Mes-Massons, A. M. (2014). Individual investigator profiles of biospecimen use in cancer research. *Biopreservation and Biobanking, 12*(3), 192–198. https://doi.org/10.1089/bio.2013.0092

11. Cole, A., Cheah, S., Dee, S., Hughes, S., & Watson, P. H. (2012). Biospecimen use correlates with emerging techniques in cancer research: Impact on planning future biobanks. *Biopreservation and Biobanking, 10*(6), 518–525. https://doi.org/10.1089/bio.2012.0038

12. Hughes, S. E., Barnes, O. R., & Watson, P. H. (2010). Biospecimen use in cancer research over two decades. *Biopreservation and Biobanking, 8*(2), 89–97. https://doi.org/10.1089/bio.2010.0005

13. Hewitt, R. E. (2011). Biobanking_ the foundation of personalized medicine. *Current Opinion in Oncology, 23*(1), 112–119. https://doi.org/10.1097/CCO.0b013e32834161b8

14. Bevilacqua, G., Bosman, F., Dassesse, T., Höfler, H., Janin, A., Langer, R., Larsimont, D., Morente, M. M., Riedman, P., Schirmacher, P., Stanta, G., Zatloukal, K., Caboux, E., & Hainaut, P. (2010). The role of the pathologist in tissue banking: European consensus expert group report. *Virchows Archiv, 456*(4), 449–454. https://doi.org/10.1007/s00428-010-0887-7

15. Guerin, J. S., Murray, D. W., McGrath, M. M., Yuille, M. A., McPartlin, J. M., & Doran, P. P. (2010). Molecular medicine Ireland guidelines for standardized biobanking. *Biopreservation and Biobanking, 8*(1), 3–63. https://doi.org/10.1089/bio.2010.8101

16. Chandrasekar, A., Warwick, R. M., & Clarkson, A. (2011). Exclusion of deceased donors post-procurement of tissues. *Cell and Tissue Banking, 12*(3), 191–198. https://doi.org/10.1007/s10561-010-9184-6

17. CNIO Stop Cancer. (2021). *The CNIO Biobank.* Retrieved March 12, 2021, from http://www.cnio.es/ing/grupos/plantillas/presentacion.asp?grupo=50004308

18. Wales Cancer Bank. (2021). Retrieved March 12, 2021, from http://www.walescancerbank.com

19. Campbell, L. D., Astrin, J. J., DeSouza, Y., Giri, J., Patel, A. A., Rayley-Payne, M., Rush, A., & Sieffert, N. (2018). The 2018 revision of the ISBER best practices: Summary of changes and the editiorial team's development process. *Biopreservation and Biobanking, 16*(1), 3–6. https://doi.org/10.1089/bio.2018.0001

20. Expert Group on Dealing with Ethical and Regulatory Challenges of International Biobank Research. (2012). *Biobanks for Europe: A challenge for governance*. Retrieved March 12, 2021, from https://www.coe.int/t/dg3/healthbioethic/Activities/10_Biobanks/biobanks_for_Europe.pdf

21. BBMRI-ERIC. (2021). Retrieved March 12, 2021, from http://www.bbmri-eric.eu/

22. Kauffmann, F., & Cambon-Thomsen, A. (2008). Tracing biological collections: Between books and clinical trials. *JAMA, 299*(19), 2316–2318. https://doi.org/10.1001/jama.299.19.2316

23. Artene, S. A., Ciurea, M. E., Purcaru, S. O., Tache, D. E., Tataranu, L. G., Lupu, M., & Dricu, A. (2013). Biobanking in a constantly developing medical world. *The Scientific World Journal, 2013*, 343274.

Types of Biobanks

3

Svetlana Gramatiuk and Berthold Huppertz

Abstract

As already introduced in Chap. 1, a biobank is a type of biorepository where biological samples, usually of human nature, are stored for research. Like described before, the term "biobank" can be defined as a collection of human biological samples and related data. These samples and data are systematically organized for research purposes. While the collection of biospecimens from other living organisms can also be called a biobank, many prefer to use this term only for human biospecimens. Both the terms "biobanks" and "biorepositories" are conceptually compatible. In this chapter, you will learn more about different classifications of biobanks—for example, biobanks can be classified based on different approaches, they can also differ in scale, quality, subject matter, and participants, and some collections can be defined by donor arrangement.

Keywords

Classification of biobanks · Biobank versus biorepository · Population-based · Disease-oriented · Institution- or clinic-based · Biobanks of network projects

At the first stage of introducing and comparing different classifications and types of biobanks, it is necessary to clarify that, from the point of view of biobanks and biorepositories, "biobank" and "biorepository" are compatible by concepts [1–5]. The term "center of biological resources" was used for "biobanks" when

S. Gramatiuk
Institute of Cellular Biorehabilitation, Kharkiv, Ukraine

B. Huppertz (✉)
Division of Cell Biology, Histology and Embryology, Gottfried Schatz Research Center, Medical University of Graz, Graz, Austria
e-mail: berthold.huppertz@medunigraz.at

conducting cancer research. Additionally, biobanks are also providers of medical research services in general; thus, "biobanks" are not only collecting human biological material, but the term is used in a much wider sense [1, 2, 4–8].

The National Cancer Institute in the United States of America has described the often-used definition "biorepository" as an especially organized abode or building in which biological substances, under certain temperature conditions, are stored [9, 10]. With the same meaning, the term "biobank" was used by further American and European organizations [10, 11].

A taxonomy of types of biobanks can be made if the systematic interaction of a biobank with its participants is the basis of the differentiation. Among others, there are differently typified categories of biorepositories like *population-based, disease-oriented, institution- or clinic-based, biobanks of network projects, moved along by the course of the host organizations, or national public ones* [12–17].

Biobanks also differ in scale, quality, subject matter, and participants. Some definitions refer to the ground stage and stage of development of biobanks. There can be *"radically converse opinions over how fixed reputation should be recognized, formulated, explained, or high-spirited..."* [13, 18, 19].

Human biorepositories have a specific classification [20]:

- *Tissue type*: (tumor and/or other benign tissues, blood, RNA, or DNA, etc.).
- *Topic*: Planned application or purpose (scientific investigation, therapy like transplantation (may be also forensics) can be the origin for therapeutic procedures such as umbilical material, stem cell, also investigation or monitoring of therapy, etc.) [7, 8, 14].
- *Ownership*: Universities or other academic institutions and academic groups, clinics, pharmaceutical companies as well as biotechnology industries, biobanking networks, or societies (e.g., rare disease biobanks and all others that have funds, to be able to sustain a biobank). Ownership can be in co-partnership, public but also private, through area and field borders (proprietorship is sometimes general, or based on performance in company).

The collection can also be defined by donor arrangement (inhabitants: adults, pregnant women, sick people, newborn, contain limited types and diagnosis) [19]. They can also be differentiated by the size of collection spread (condition based on one clinic, cross-sectional in public, state-wide, or in one city).

Gottweis and Zatloukal [16] distinguish four categories of human biobanks:

- *Clinical case-controlled biobanks* built up with human material of individuals having given diagnosis and correspondingly healthy controls. A nice example is the pathology archive. These kinds of "biobanks" are confronted with huge challenges, as archives such as those initiated in pathologies often do not follow strict ethical guidelines for scientific usage (e.g., ethical approval or informed consent).
- *Longitudinal population-based biobanks*, which monitor a concrete and defined part of a given group of inhabitants living in a specific region for a long time (e.g., *the Estonian and UK Biobank*).

- *Isolated population biobanks*, which are characterized by a consistent genetical basis as well as ecological circumstances for a target population (e.g., *the Icelandic Biobank*).
- *Twin registries* are characterized as a collection of human biological material derived from twins (independently if dizygotic or monozygotic) (e.g., *the GenomEUtwin and the Swedish Twin registry*).

Rebulla et al. [21, 22] make the categorization of biobanks more advanced. They distinguish between six classified types of biobanks:

- *"Leftover tissue biobanks" collected during clinical pathology diagnostic procedures*
- *Population biobanks*
- *Twin biobanks*
- *Disease biobanks from patients suffering specific conditions*
- *Organ biobanks*
- *Nonhuman biobanks (e.g., "Primate Brain Bank")*

Currently, a comprehensively acknowledged classification is introduced by BBMRI [19, 23, 24] that differentiates between two specific categories of biorepositories:

- *Population-based biobanks* (population-based biorepositories, which are prospectively oriented on the observation (survey) of the usual way of life and detection of diseases and/or their complications over a long period)
- *Disease-oriented biobanks* (human biological samples such as tissue samples and corresponding diagnostic data are usually available for biomedical research that are more complex and clinically relevant)

References

1. Organization for Economic Cooperation and Development. (2007). *OECD best practices guidelines for biological resource centers.* Retrieved from http://www.oecd.org/sti/emerging-tech/oecdbestpracticeguidelinesforbiologicalresourcecentres.htm
2. Vaught, J., Rogers, J., Carolin, T., & Compton, C. (2011). Biobankonomics: Developing a sustainable business model approach for the formation of a human tissue biobank. *JNCI Monographs, 2011*(42), 24–31. https://doi.org/10.1093/jncimonographs/lgr009
3. Gaffney, E. F., Madden, D., & Thomas, G. A. (2011). The human side of cancer biobanking. *Molecular Profiling, 2012*(823), 59–77. https://doi.org/10.1007/978-1-60327-216-2_5
4. OECD. (2014, January). *Emerging technologies. Guidelines for human biobanks and genetic research databases (HBGRDs).* Retrieved from http://www.oecd.org/sti/emerging-tech/guidelines-for-human-biobanks-and-genetic-research-databases.htm
5. OECD Organization for Economic Cooperation and Development. (2010). OECD guidelines on human biobanks and genetic research databases. *European Journal of Health Law, 17*(2), 191–204.

6. National Cancer Institute, National Institutes of Health, U.S. Department of Health and Human Services. (2007, June). *National Cancer Institute best practices for biospecimen resources.* Retrieved from https://biospecimens.cancer.gov/bestpractices/2007-NCIBestPractices.pdf
7. Botti, G., Franco, R., Cantile, M., Ciliberto, G., & Ascierto, P. A. (2012). Tumor biobanks in translational medicine. *Journal of Translational Medicine, 10*(204). https://doi.org/10.1186/1479-5876-10-204
8. Hanif, Z., Sufiyan, N., Patel, M., & Akhtar, M. Z. (2018). Role of biobanks in transplantation. *Annals of Medicine and Surgery, 28*, 30–33. https://doi.org/10.1016/j.amsu.2018.02.007
9. Colledge, F., Elger, B., & Howard, H. C. (2013). A review of the barriers to sharing in biobanking. *Biopreservation and Biobanking, 11*(6), 339–346. https://doi.org/10.1089/bio.2013.0039
10. National Cancer Institute. (2016). *NCI Best practices for biospecimen resources.* Retrieved July 22, 2018, from biospecimens.cancer.gov/bestpractices/2016-NCIBestPractices.pdf
11. Campbell, L. D., Astrin, J. J., DeSouza, Y., Giri, J., Patel, A. A., Rayley-Payne, M., Rush, A., & Sieffert, N. (2018). The 2018 revision of the ISBER best practices: Summary of changes and the editorial Team's development process. *Biopreservation and Biobanking, 16*(1), 3–6. https://doi.org/10.1089/bio.2018.0001
12. Colledge, F., Persson, K., Elger, B., & Shaw, D. (2013). Sample and data sharing barriers in biobanking: Consent, committees, and compromises. *Annals of Diagnostic Pathology, 18*(2), 78–81. https://doi.org/10.1016/j.anndiagpath.2013.12.002
13. Fortin, S., Pathmasiri, S., Grintuch, R., & Deschênes, M. (2011). 'Access arrangements' for biobanks: A fine line between facilitating and hindering collaboration. *Public Health Genomics, 14*(2), 104–114. https://doi.org/10.1159/000309852
14. Luque, J. S., Quinn, G. P., Montel-Ishino, F. A., Arevalo, M., Bynum, S. A., Noel-Thomas, S., Wells, K. J., Gwede, C. K., Meade, C. D., & Tampa Bay Community Cancer Network Partners. (2012). Formative research on perceptions of biobanking: What community members think. *Journal of Cancer Education, 27*(1), 91–99. https://doi.org/10.1007/s13187-011-0275-2
15. Mascalzoni, D., Dove, E. S., Rubinstein, Y., Dawkins, H. J., Kole, A., McCormack, P., Woods, S., Riess, O., Schaefer, F., Lochmüller, H., Knoppers, B. M., & Hansson, M. (2016). International charter of principles for sharing biospecimens and data. *European Journal of Human Genetics, 24*(7), 1096. https://doi.org/10.1038/ejhg.2015.237
16. Gottweis, H., & Zatloukal, K. (2007). Biobank governance: Trends and perspectives. *Pathobiology, 74*(4), 206–211. https://doi.org/10.1159/000104446
17. Downey, P., & Peakman, T. C. (2008). Design and implementation of a high-throughput biological sample processing facility using modern manufacturing principles. *International Journal of Epidemiology, 37*(1), i46–i50. https://doi.org/10.1093/ije/dyn031
18. Brisson, A. R., Matsui, D., Rieder, M. J., & Fraser, D. D. (2012). Translational research in pediatrics: Tissue sampling and biobanking. *Pediatrics, 129*(1), 153–162. https://doi.org/10.1542/peds.2011-0134
19. Friede, A., Grossman, R., Hunt, R., Li, R. M., & Stern, S. (Eds.). (2003). *National Biospecimen Network Blueprint.* Constella Group.
20. BioMedInvo4All. (2018). Retrieved July 22, 2018, from www.biomedinvo4all.com/en/research-themes/medical-data-and-biobanks/medical-data-and-biobanks-basics
21. Rebulla, P., Lecchi, L., Giovanelli, S., Butti, B., & Salvaterra, E. (2007). Biobanking in the year 2007. *Transfusion Medicine and Hemotherapy, 34*, 296–292. https://doi.org/10.1159/000103922
22. Albert, M., Bartlett, J., Johnston, R. N., Schacter, B., & Watson, P. (2014). Biobank bootstrapping: Is biobank sustainability possible through cost recovery? *Biopreservation and Biobanking, 12*(6), 374–380. https://doi.org/10.1089/bio.2014.0051
23. BBMRI-ERIC. (2021). Retrieved March 12, 2021, from http://www.bbmri-eric.eu/
24. CORDIS. (2015). *Biobanking and biomolecular resources research infrastructure.* Retrieved July 22, 2018, from https://cordis.europa.eu/result/rcn/162431_de.html

Ethical and Legal Principles in the Field of Biobanking

4

Svetlana Gramatiuk, Tanja Macheiner, Christine Mitchell, and Karine Sargsyan

Abstract

The trend to store biological material with associated data in biobanks has increased concerns on how to protect ethics in the handling of human biological samples and associated data. The invasive collection of body substances for biobanks is a complex issue in biobank ethics and is widely discussed. By giving an informed consent for a medical research approach, a person gives their acceptance to a researcher for scientific investigations on the person's samples and related data. This chapter highlights the impact of biomedical research and the frames of ethical action and gives information on the most frequent guidelines and regulations on regulatory, legal, and ethical aspects of biobanking.

Keywords

Informed consent · Data · Ethics · Biobanking law · Data protection · Protection rules · ELSI

The independence guidelines ensure that each individual can decide for himself/herself to consent to a biobank collection in which his/her personal data and samples

S. Gramatiuk
Institute of Cellular Biorehabilitation, Kharkiv, Ukraine

T. Macheiner · C. Mitchell
International Biobanking and Education, Medical University of Graz, Graz, Austria

K. Sargsyan (✉)
International Biobanking and Education, Medical University of Graz, Graz, Austria

Department of Medical Genetics, Yerevan State Medical University, Yerevan, Armenia

Ministry of Health of the Republic of Armenia, Yerevan, Armenia
e-mail: karine.sargsyan@medunigraz.at

are stored, possibly influencing him/her. The invasive collection of body substances for biobanks (albeit minimal) is also a complex issue in biobank ethics and is widely discussed. Self-expression involves the direct determination of the application to which one's own physical samples and data is placed (information specific conclusion). Any restriction on equal rights of a particular person must also be protected in a specific way [1–6].

By means of an informed consent to a medical research approach, a person is giving an acceptance to a researcher for scientific investigations on the person's samples and related data. Medical, clinical, and ethical societies have communicated a reasonable number of settings and instructions proposed for the protection of a person's right to independence. It is important that consent to medical research must be voluntary and clearly defined at all times, directly based on the full evidence available to the researcher at the moment of consent. The rules precisely describe that consenting must be professional and has to fully inform the donor/volunteer about the nature, the whole scope, and the possible risks of the sampling on his/her life, health, and body and, if already known, about any objectives of the research [1, 2, 4–6]. Nevertheless, the described principles must be proven on applicability in specific settings of the given biobanks.

Although the contributors' autonomy as an essential validity characteristic forms the basis both for the establishment and for the development of the use of biobanks, it is not the only rule and is not sufficient. No single person can raise his/her independence for justification of activities by which the rights of another person or general public entity can be overstepped. Therefore, regardless of the valid informed consent of the donor, there is a general question raised by ethics specialists: If biorepositories carry undesirable risks for third parties and/or the general public—for example, leading to stigmatization or discrimination—then respective guidelines need to be in place to hinder such processes [7–11].

All medical doctors in their profession are asked, based on professional ethics, to promote the safety and well-being of their patients. This fact is not intended to limit the medical doctor-patient relationship in connection with diagnostic or therapeutic measures but is intended for research purposes and may even be extended to potential donors. In this way, the implementation of medical science can be pushed forward, and this can also benefit potential future patients. What can be expected from a medical doctor (who can be at the same time the researcher), if not a professional obligation, which is something that can be justifiably and responsibly anticipated from a medical professional? Physicians who request contributors to provide human biological material and corresponding medical-clinical information for biorepositories are usually serving needs of biomedical science: This means he/she is not simply following his/her professional benefits as a curing specialist; he/she correspondingly desires to make a confident moral contribution supporting medical research for future innovation. This is an additional facet of consideration of the legitimacy of such donations [7, 10, 11].

Biobanks in general are not only questioned for their perspective of avoidance of probable dangers and risks but also concerning their specific and societal usefulness. The awareness of the utility of biorepositories and biobanks in general may also be necessary by reflections of public solidarity. Many donors regard their readiness to provide their human biological material and related information as an expression of their own responsibility to support other people and future generations. The confident moral commitment to support the public in general is an essential element and is in contrast to the reverse moral responsibility of "primum non nocere"—which means "first of all do no harm" [10–13].

Current "data protection" legislation allows consideration of the use of individual and/or clinical information, which is permitted in the absence of specific donor consent. On the other hand, the requirements of these legal and ethical regulations are generally unperturbed in support of biomedical research. For example, both public and nonpublic bodies are allowed to save data, which can be subject of particular protection rules: More specifically, it can be information on health, sexual orientation, or ethnic origin. These parameters can be stored in biobanks or study cohort databases even without the donors' explicit and informed consent, only if the scientific interest and expected results from specific research activities substantially overshadow the single interest of those donors. As a rule, these data may only be used if the objective of the study cannot be achieved by other means and/or can only be achieved at very high cost (this is mostly discussed when it comes to public money). Moreover, public bodies are in many cases obliged to use information already collected for their stated purposes. Once a valuable set of information has been collected at a very high effort and investment, it cannot be moral to collect again at twice the expense. Thus, if information is collected for the use in scientific studies (own collection) but can be passed on anonymously to public or nonpublic bodies (for a fee and for a higher goal of the health and well-being of the population), in this case the interest in the project and the interest of the parties involved, who may be concerned that the subject of the data use "does not change," may far outweigh the interest in the project.

Occasionally, the aim of scientific investigations may be not accomplished in any other way as to use the given dataset, and it can be of high interest and benefit to the human population. Finally, it can be subject to approximately the same circumstances, when in a country (such as the Ukraine) nonpublic bodies are allowed to consume even private data (if specifically administered by and in authority) even if they do not correspond to the initial purpose.

In addition to the Data Protection Law, some specific regions (states and/or provinces) may have the same data protection legalizations valid for the given area and definite information protection commandments as well as guidelines related to clinical settings and general health segments. This will affect studies on a larger or smaller level. These laws and regulations fluctuate significantly in their necessities. A differentiation for the recognition of scientific investigations is often made when they are carried out in the own research institution and/or when they are carried out outside the respective health institution. Often the mentioned legal mechanisms deliver acceptability of personal and clinical information usage, but still can be adapted by an evaluation of goals and benefits (public benefit for conducting medical scientific investigations required to overshadow, and/or considerably be more important than person's interests). Several circumstances indicate that an independent entity must be included in the process to ensure that there are no inconsistencies regarding protection issues that may have a profound impact on the health and well-being of the research participant or that any agreed-upon benefits are not overridden. A conclusive principle is often that instead of personalized information, there should be no other unconventional variances but that every use, regardless of whether it is anonymized or not, must be documented and known. In conclusion, several biomedical tools involve data usage authorization by specialized bodies. Full adoption demonstrates the need to implement, maintain, and—if possible—improve an adequate privacy and data security law, regardless of its potentially enormous complex and inconsistent presentation.

References

1. Guerin, J. S., Murray, D. W., McGrath, M. M., Yuille, M. A., McPartlin, J. M., & Doran, P. P. (2010). Molecular medicine Ireland guidelines for standardized biobanking. *Biopreservation and Biobanking, 8*(1), 3–63. https://doi.org/10.1089/bio.2010.8101
2. Chandrasekar, A., Warwick, R. M., & Clarkson, A. (2011). Exclusion of deceased donors post-procurement of tissues. *Cell and Tissue Banking, 12*(3), 191–198. https://doi.org/10.1007/s10561-010-9184-6
3. Expert Group on Dealing with Ethical and Regulatory Challenges of International Biobank Research. (2012). *Biobanks for Europe: A challenge for governance*. Retrieved March 12, 2021, from https://www.coe.int/t/dg3/healthbioethic/Activities/10_Biobanks/biobanks_for_Europe.pdf
4. Campbell, B., Thomson, H., Slater, J., Coward, C., Wyatt, K., & Sweeney, K. (2007). Extracting information from hospital records: What patients think about consent. *Quality and Safety in Health Care, 16*(6), 404–408. https://doi.org/10.1136/qshc.2006.020313
5. Sándor, J., Bárd, P., Tamburrini, C., & Tännsjö, T. (2012). The case of biobank with the law: Between a legal and scientific fiction. *Journal of Medical Ethics, 38*(6), 347–350. https://doi.org/10.1136/jme.2010.041632
6. Knoppers, B. M. (2005). Biobanking: International norms. *The Journal of Law. Medicine and Ethics.* https://doi.org/10.1111/j.1748-720X.2005.tb00205.x
7. Bevilacqua, G., Bosman, F., Dassesse, T., Höfler, H., Janin, A., Langer, R., Larsimont, D., Morente, M. M., Riedman, P., Schirmacher, P., Stanta, G., Zatloukal, K., Caboux, E., & Hainaut, P. (2010). The role of the pathologist in tissue banking: European consensus expert group report. *Virchows Archiv, 456*(4), 449–454. https://doi.org/10.1007/s00428-010-0887-7
8. CNIO Stop Cancer. (2021). *The CNIO Biobank*. Retrieved March 12, 2021, from http://www.cnio.es/ing/grupos/plantillas/presentacion.asp?grupo=50004308
9. Wales Cancer Bank. (2021). Retrieved March 12, 2021, from http://www.walescancerbank.com
10. Gaffney, E. F., Madden, D., & Thomas, G. A. (2011). The human side of cancer biobanking. *Molecular Profiling, 2012*(823), 59–77. https://doi.org/10.1007/978-1-60327-216-2_5
11. Moe, J. L., Pappas, G., & Murray, A. (2017). Transformational leadership, transnational culture and political competence in globalizing health care services: A case study of Jordan's King Hussein Cancer center. *Globalization and Health, 3*(11). https://doi.org/10.1186/1744-8603-3-11
12. Knoppers, B. M., Fortier, I., Legault, D., & Burton, P. (2008). Population genomics: The public population project in genomics (P^{3}G): A proof of concept? *European Journal of Human Genetics, 16*, 664–665. https://doi.org/10.1038/ejhg.2008.55

13. German Ethics Council. (2010). *Human biobanks for research: Opinion.* Retrieved from https://www.ethikrat.org/fileadmin/Publikationen/Stellungnahmen/englisch/DER_StnBiob_ Engl_Online_mitKennwort.pdf

Business Planning for Biobanks

Svetlana Gramatiuk, Tanja Macheiner, Tamara Sarkisian, Armen Muradyan, and Gariele Hartl

Abstract

Fundamentally, scientists have the necessary knowledge to operate a biobank, but the practical application of economic tools is not sufficient in some cases. This chapter shows the importance of using economic tools for the organisation and sustainability of biobanks. It also explains the implementation of such economic tools that are essential for the correct operation of biobanking infrastructures. Furthermore, this chapter depicts features of a business plan and provides a useful basic policy to its development for biobanks, scientists, managers and stakeholders. For example, a SWOT (strengths, weaknesses, opportunities and threats) analysis is an integral part of every business plan. In addition, the definition of a mission is essential to any business plan.

Keywords

Business plan · Economic tools · Management tools · SWOT analysis · Mission statement

S. Gramatiuk
Institute of Cellular Biorehabilitation, Kharkiv, Ukraine

T. Macheiner · G. Hartl (✉)
International Biobanking and Education, Medical University of Graz, Graz, Austria
e-mail: gabriele.hartl@medunigraz.at

T. Sarkisian
Department of Medical Genetics, Yerevan State Medical University, Yerevan, Armenia

A. Muradyan
Yerevan State Medical University, Yerevan, Armenia

K. Sargsyan et al. (eds.), *Biobanks in Low- and Middle-Income Countries: Relevance, Setup and Management*, https://doi.org/10.1007/978-3-030-87637-1_5

Fundamentally, scientists have the essential knowledge to run a biobank, but the practical application of economic tools in some cases is not sufficient. Usually, the management of public healthcare facilities consists primarily of scientists, and is not supported by economists. In many academic cases, there is a broad-spectrum absence of educational background necessary for profitable planning and the usage of suitable tools aimed to manage sustainably. Therefore, this chapter aims at increasing understanding of the significance of appropriate tools for the economic and sustainable management of biobanks and at the implementation of business processes, important for efficient management of research infrastructure facilities [1–5].

SWOT Analysis

This chapter aims at describing the characteristics of a business plan and to specify a useful essential policy to its progress for biobanks, scientists, managers and stakeholders. Although this book primarily focuses on human biological material and associated biobanks, the opinions and concepts stated may be functional to any other research infrastructure or organisation for a systematic storage of samples and corresponding data for biological and biomedical science purposes [6–9].

For a biorepository or biobank, the narrative explanation of research goals in concrete projects outlines the specific characteristics of the human biological material collected and stored, together with the aim and purpose of the collection and the preservation methods used. Products of a biorepository can comprise of data, samples and the usage of associated analysis platforms (new data generation) and services, offered for any type of prospective method or implementation measures by the biobank. It is appropriate to outline the origin of biomaterial, such as infectious microorganisms, tissue or derivative of the biomaterial (body fluids, cells, nucleic acids, proteins, etc.), in addition to the type and method of specimen handling (e.g.—frozen) and storage conditions (temperature, container, quantity for aliquot, etc.). Furthermore, it is usable to include a short history of the biobank [10, 11].

For biobanks, which have been around for a long time and have a long follow-up, the biobank's past can be multifaceted depending on the inconsistent and mixed origin of the samples and the progress of the organisation. If diverse cohorts are collected in the biobank, the narrative in this segment of the business plan will comprise a completed and ongoing catalogue of stored sample collections. The administrative responsibility and functional roles among the senior staff of the biobank and the respective researchers for different collections must be stated. Forthcoming collections and follow-ups will be clearly specified. It may also be useful in this section to include a table listing the biological sample collections managed by the biobank. This table should indicate the name; the application area; the date of starting collecting; the factual development step of the project (conception, collection ongoing, completed); the socio-ethical, research and legal personnel in charge of each collection; and the specific role in relation to the senior manager of the infrastructure [3–5, 12, 13].

The SWOT analysis is an essential part of every business plan [14, 15], and it is designed to evaluate *strengths (S), weaknesses (W), opportunities (O) and threats (T)* of a project. While the first two elements are defined as intramural influences, the concluding two features are intended to be extramural factors.

The strengths of a biorepository or a biobank depend on quality standards, the amount of collected biological material and the categories of human samples and simply standard processes for sample usage in research projects. Quality standards mean, for example, fulfilment of specific quality requirements and techniques, as well as possible certification. These quality standards and variety of the data connected to the samples are a strength for the biorepository or a specific collection in it. Biobanks that work on the basis of an internationally recognised quality management system (QMS) are preferred associates for the elaboration of new products. *The International Organization for Standardization (ISO)* is currently determining the obligations for bioresources in research and market application together with the working group on biobanking created at Technical Committee (TC) 276 Biotechnology [5, 16–19].

Important elements like clinical origin, precise data on the pre-analytical procedures, the study size, the geographical origin, the inclusion criteria of an organisation with a high operating force, the trustworthiness of the organisation, the certification, the dimension of obtainable information connected with the samples, the extraordinary products, the laboratory methods and infrastructure and the sample storage systems are significant for the exclusivity or the uncommonness of samples. More strengths originate from the competency to deliver efficient services (e.g. a user-friendly and well-defined policy of sample usage, reasonable costs for usage, effective supply schemes, processing rapidity and the safety of data transfer).

One of the frequently documented weaknesses of biobanks is documented to be the struggle to implement a total cost recovery model, especially in the case of high infrastructure costs, independent on the biobank's implemented economic model and on number of samples issued. In addition, a weakness can be stated by biobanking in projects with an expiry date. In these cases, the future of human biological samples collected within the scope of the given funded project is undecided, therefore open [7–10, 15, 20, 21].

Opportunities can raise if an option of development and/or installation of a central biobank facility (institution-wide, region-wide, national, global) as a reference centre for sample management in a particular field of activities is offered. Also, worth mentioning are the opportunities to become member of professional networks and to collaborate with other centres to improve the availability of human biological material. It is also important to consider possible additional research services that can be offered to academic and industrial areas.

Threats: Disasters and the missing of backup plans are one of the key risks to which biobanks could be exposed [7–9, 18, 22, 23].

Different kinds of accidents (e.g. natural, human or technological) may threaten a biobank. In the SWOT analysis of a biobank, it is advisable to add to the part of threats a section, which is devoted to alternative possibilities and the strategies of the variance development. In these documents, all added costs, especially for the risk assessment, preliminary plans of mitigation in case of emergency and contingency plans, are needed to be developed, documented and well thought through specifically adjusted for the place and ecosystem of the specified biobank [8–10].

Economic malfunction results from a reduction of public funding or charitable support, insolvency or the termination of a host organisation. Changes in accepted organisational concepts for biobanks, the loss of project-oriented financing and other possible financial changes could represent a source of risk for a biobank.

Further threats can come from ethical-legal issues, which can contain enormous withdrawal rates of consent due to the public impact of a non-existent news item, which can be a source of sudden mistrust in biobanks.

The Mission

A mission statement (e.g. goals and aims of the organisation) is crucial for a business plan. Mission, often outlined in very short message, explains the presence of the company and directs the services of the organisation. While the mission is an easy-to-understand area of a business plan, thoughtful conception is greatly suggested, both for companies and biobanks [9, 10, 24].

The mission statement targets to identify the main capacity of a biobank, and it shows wherefore biobanking, even when implemented in a truly wide-ranging organisation, has been implemented, exists and must be maintained.

Overall, the general project outline and mission statement have a duty to describe and present the tasks and duties aligned with a narrative of the fundamental ideas and history of the biobank supplemented with comprehensive explanation of samples, sample collections, research services, working guidelines and workforces' relations.

SWOT analysis and substitute policies are the essential fragment of this segment. Taking into relation the amount and significance of the stated data, the use of graphical presentation is extremely suggested.

References

1. Knoppers, B. M., Fortier, I., Legault, D., & Burton, P. (2008). Population genomics: The public population project in genomics (P³G): A proof of concept? *European Journal of Human Genetics, 16*, 664–665. https://doi.org/10.1038/ejhg.2008.55
2. Yu, X., Bian, X., Throop, A., Song, L., Del Moral, L., Park, J., Seiler, C., Fiacco, M., Steel, J., Hunter, P., Saul, J., Wang, J., Qiu, J., Pipas, J. M., & LaBaer, J. (2014). Exploration of panviral proteome: High-throughput cloning and functional implications in virus-host interactions. *Theranostics, 4*(8), 808–822. https://doi.org/10.7150/thno.8255
3. Brat, D. J., Verhaak, R. G., Aldape, K. H., Yung, W. K. A., Salama, S. R., Cooper, L. A. D., Rheinbay, E., Miller, C. R., Vitucci, M., Morozova, O., Robertson, A. G., Noushmehr, H., Laird, P. W., Cherniack, A. D., Akbani, R., Huse, J. T., Ciriello, G., Poisson, L. M., Barnholtz-Sloan, J. S., ... Zhang, J. (2015). Comprehensive, integrative genomic analysis of diffuse lower-grade gliomas. *New England Journal of Medicine, 372*(26), 2481–2498. https://doi.org/10.1056/NEJMoa1402121

4. Colledge, F., Elger, B., & Howard, H. C. (2013). A review of the barriers to sharing in biobanking. *Biopreservation and Biobanking, 11*(6), 339–346. https://doi.org/10.1089/bio.2013.0039

5. Colledge, F., Persson, K., Elger, B., & Shaw, D. (2013). Sample and data sharing barriers in biobanking: Consent, committees, and compromises. *Annals of Diagnostic Pathology, 18*(2), 78–81. https://doi.org/10.1016/j.anndiagpath.2013.12.002

6. Downey, P., & Peakman, T. C. (2008). Design and implementation of a high-throughput biological sample processing facility using modern manufacturing principles. *International Journal of Epidemiology, 37*(1), i46–i50. https://doi.org/10.1093/ije/dyn031

7. Albert, M., Bartlett, J., Johnston, R. N., Schacter, B., & Watson, P. (2014). Biobank bootstrapping: Is biobank sustainability possible through cost recovery? *Biopreservation and Biobanking, 12*(6), 374–380. https://doi.org/10.1089/bio.2014.0051

8. Clément, B., Yuille, M., Zatloukal, K., Wichmann, H. E., Anton, G., Parodi, B., Kozera, L., Bréchot, C., Hofman, P., & Dagher, G. (2014). EU-US Expert Group on cost recovery in biobanks. *Science Translational Medicine, 6*(261), 261fs45. https://doi.org/10.1126/scitranslmed.3010444

9. Warth, R., & Perren, A. (2014). Construction of a business model to assure financial Sustainability of biobanks. *Biopreservation and Biobanking, 12*(6), 389–394. https://doi.org/10.1089/bio.2014.0057

10. Vaught, J., Rogers, J., Carolin, T., & Compton, C. (2011). Biobankonomics: Developing a sustainable business model approach for the formation of a human tissue biobank. *JNCI Monographs, 2011*(42), 24–31. https://doi.org/10.1093/jncimonographs/lgr009

11. Hirtzlin, I., Dubreuil, C., Préaubert, N., & Duchier, J. (2003). An empirical survey on biobanking of human genetic material and data in six EU countries. *European Journal of Human Genetics, 11*(6), 475–488. https://doi.org/10.1038/sj.ejhg.5201007

12. Fortin, S., Pathmasiri, S., Grintuch, R., & Deschênes, M. (2011). 'Access arrangements' for biobanks: A fine line between facilitating and hindering collaboration. *Public Health Genomics, 14*(2), 104–114. https://doi.org/10.1159/000309852

13. Sándor, J., Bárd, P., Tamburrini, C., & Tännsjö, T. (2012). The case of biobank with the law: Between a legal and scientific fiction. *Journal of Medical Ethics, 38*(6), 347–350. https://doi.org/10.1136/jme.2010.041632

14. Macheiner, T., Huppertz, B., Bayer, M., & Sargsyan, K. (2017). Challenges and driving forces for business plans in biobanking. *Biopreservation and Biobanking, 15*(2), 121–125. https://doi.org/10.1089/bio.2017.0018

15. Sargsyan, K., Macheiner, T., Story, P., Strahlhofer-Augsten, M., Plattner, K., Riegler, S., Granitz, G., Bayer, M., & Huppertz, B. (2015). Sustainability in biobanking: Model of biobank Graz. *Biopreservation and Biobanking, 13*(6), 410–420. https://doi.org/10.1089/bio.2015.0087

16. Schenk, M., Huppertz, B., Obermayer-Pietsch, B., Kastelic, D., Hörmann-Kröpfl, M., & Weiss, G. (2016). Biobanking of different body fluids within the frame of IVF—A standard operating procedure to improve reproductive biology research. *Journal of Assisted Reproduction and Genetics, 34*, 383–290. https://doi.org/10.1007/s10815-016-0847-5

17. Moser, G., & Huppertz, B. (2017). Implementation and extravillous trophoblast invasion: From rare archival specimens to modern biobanking. *Placenta, 56*, 19–26. https://doi.org/10.1016/j.placenta.2017.02.007

18. Mascalzoni, D., Dove, E. S., Rubinstein, Y., Dawkins, H. J. S., Kole, A., McCormack, P., Woods, S., Riess, O., Schaefer, F., Lochmüller, H., Knoppers, B. M., & Hansson, M. (2016). International charter of principles for sharing biospecimens and data. *European Journal of Human Genetics, 24*(7), 1096. https://doi.org/10.1038/ejhg.2015.237

19. Vaught, J. B. (2006). Biorepository and biospecimen science: A new focus for CEBP. *Cancer Epidemiology, Biomarkers & Prevention, 15*(9), 1572–1573. https://doi.org/10.1158/1055-9965.EPI-06-0632

20. Odeh, H., Miranda, L., Rao, A., Vaught, J., Greenman, H., McLean, J., Reed, D., Memon, S., Fombonne, B., Guan, P., & Moore, H. M. (2015). The biobank economic modeling tool

(BEMT): Online financial planning to facilitate biobank sustainability. *Biopreservation and Biobanking, 16*(6), 421–429. https://doi.org/10.1089/bio.2015.0089

21. Watson, P. H., Nussbeck, S. Y., Cater, C., O'Donoghue, S., Cheah, S., Matzke, L. A. M., Barnes, R. O., Bartlett, J., Carpenter, J., Grizzle, W. E., Johnston, R. N., Mes-Masson, A.-M., Murphy, L., Sexton, K., Shepherd, L., Simeon-Dubach, D., Zeps, N., & Schacter, B. (2014). A framework for biobank sustainability. *Biopreservation and Biobanking, 12*(1), 60–68. https://doi.org/10.1089/bio.2013.0064

22. Gottweis, H., & Zatloukal, K. (2007). Biobank governance: Trends and perspectives. *Pathobiology, 74*(4), 206–211. https://doi.org/10.1159/000104446

23. De Souza, Y. G., & Greenspan, J. S. (2014). Biobanking past, present and future: Responsibilities and benefits. *AIDS, 27*(3), 303–312. https://doi.org/10.1097/QAD.0b013e32835c1244

24. Braun, L., Lesperance, M., & Mes-Massons, A. M. (2014). Individual investigator profiles of biospecimen use in cancer research. *Biopreservation and Biobanking, 12*(3), 192–198. https://doi.org/10.1089/bio.2013.0092

Biobanking Concepts Specific for Developing Countries

6

Svetlana Gramatiuk, Mykola Alekseenko, Tamara Sarkisian, Armen Muradyan, and Karine Sargsyan

Abstract

Biobanks currently describe and report being present and working on every continent of the earth, with the highest density detected in North America and Europe. However, this development changes very quickly. Some countries, especially low- and middle-income countries, have recruited and invested money and excessive work for the construction of their own biobanks and biobanking networks. The biobanking activities in these countries are mostly based on the collaboration of large organizations, such as the African Society of Human Genetics, the National Institutes of Health in the United States and the Wellcome Trust. In this chapter, we list and investigate several problems and information of developing countries regarding the construction of biobanks.

Keywords

Biobank networks · Low- and middle-income countries · Biobanking societies

S. Gramatiuk · M. Alekseenko
Institute of Cellular Biorehabilitation, Kharkiv, Ukraine

T. Sarkisian
Department of Medical Genetics, Yerevan State Medical University, Yerevan, Armenia

A. Muradyan
Yerevan State Medical University, Yerevan, Armenia

K. Sargsyan (✉)
International Biobanking and Education, Medical University of Graz, Graz, Austria

Department of Medical Genetics, Yerevan State Medical University, Yerevan, Armenia

Ministry of Health of the Republic of Armenia, Yerevan, Armenia
e-mail: karine.sargsyan@medunigraz.at

Biobanks in developing countries have some particular features. Biobanks presently describe and report to be present and work on every continent of the earth, even in Antarctica, with the highest density detected in *North America and Europe* [1, 2]. However, this spreading pattern is shifting promptly. Several countries, including *China, Gambia, Jordan, Mexico and South Africa*, and many others have recruited invested money and excessive work for the construction of their own biobanks and biobanking networks [2–7].

The above-mentioned countries have several partners among biorepository facilities in high-income countries. One of the known initiatives is the consolidated *Gambian National DNA Bank*, which has been implemented based on the support of the *Centre d'Etude du Polymorphisme Humain*, which is an international scientific centre for genetics situated in *Paris, France*. As a known example, one can also mention the *Kadoorie Study of Chronic Disease in China* [8] and the *Mexico City Prospective Study*, which work together with *Oxford's Clinical Trial Service Unit and Epidemiological Studies Unit* [8]. There are several other examples worth mentioning, such as the *KHCCBIO project in Jordan* which intends to collect and store cancer samples from all over the country and also has collaborations with the *Trinity College in Dublin, Biostór Ireland and Accelopment AG, Switzerland* [9]. Also well spread and known is the initiative called *Human Heredity and Health in Africa (H3Africa)* [10]. This biobanking activity is built with a cooperation of three organizations, the *African Society of Human Genetics*, the *National Institutes of Health in the United States* and the *Wellcome Trust* [10].

The creation of biorepositories is the essential phase on the road to launching a national genomics research programme. Nevertheless, the expansion of many different biobanks is facing difficulties. Sustaining these biobanks and generating effective research results, which are based on the systematic and organized biobanking resources, can become tricky, especially without an appropriate framework and dedicated management capacity. Additionally, several countries with specific political regimens—for example, China and South Africa—have deficiency in acceptable legislative structures and regulations that may standardize or control the usage and progress of biobanks [9, 11–15].

In high-income countries, biobanks and their organizations support scientists to carry out human and especially genetic research. This can be an additional benefit, as human biological material can be studied from whole populations, especially those known to be rich in genetic diversity compared to low- and middle-income countries (LMICs). For these investigations, human biological material can be transported from a LMIC biorepository from the above-mentioned institutes, and researchers can have a secondment in the given biorepository [16]. This human biological material can cause difficulties in LMICs, as the majority of these countries apply insufficient and/or non-existent regulations and laws to protect donors. The mentioned deficiency in jurisdictive frameworks can result in vulnerability of the country and its population to exploitation [17].

The Washington Post printed (in December 2000) a six-part sequence of articles with the title: "The body hunters that surveyed research subjects in China, Africa and Latin America". The population was used as a research subject, yet people complained, that as partaking in biomedical research, which was led by high-income country scientists, they did not obtain the promised and estimated benefits—for instance healthcare services [18–21]. There are several further stories of investigators coming from developed countries, who were gathering human biological materials from *Hagahai* individuals in *Papua New Guinea*, *Havasupai* people in *Arizona* and *Karitiana* folk in *Brazil* not having any or appropriate informed consents [22]. The contributors stated their disappointment, as they did not obtain any of the expected benefits, which should have been financial remuneration and medicines.

In 2002 in India, a strict governmental regulation against biopiracy was issued, but it is still poorly realized, and human biological specimens are quietly being distributed abroad for research lacking a proper ethical and authority approval [11]. There is a very interesting systematic review on existing human genetic projects including DNA specimens from the Cameroon population, which were conducted between 1989 and 2009. This review reports of only 14% of the organizations coming from Cameroon and only 28% of Cameroonian authors, who were somehow related to the identified 50 articles. Additionally, only a few of the published research articles have mentioned topics that focus on Cameroon and the common genetic diseases in Africa. The worst result, however, showed that almost all the DNA samples of the Cameroonian population were stored far away from Africa [9, 12–14].

Scientists can indeed receive financial repayments, but also individual recognition and standing, by giving industrial partners access to biorepositories and—even worse—by commercializing biobank resources without considering and/or ignoring that this may potentially harm the welfare of donors. Biased profit sharing with research participants (donors) of an area and/or with entire populations can cause exploitation. This kind of acting leads to a population-wide distrust in biomedical research. Additionally, lacking or partly performed consent procedures and insufficient commitment (individual and organizational) cover up the relationship between science and the public [17, 18, 23, 24].

Scientific or financial profit-distribution questions due to human biological material and associated information movements across borders have been actively discussed. Numerous investigations have deliberated and provided suggestions for possible non-discriminatory profit distributions of genetic investigation partnerships across different nations [17–21, 23–25]. Nonetheless, this very essential matter and its theoretical and practical complexity remain unresolved. Specific ethical and legal strategies and guidelines have been developed and issued by several international organizations for the access to samples and the corresponding information, such as the *Human Genome Organization Ethics Committee's Statement on Benefit Sharing (2000)*, the *United Nations Educational, Scientific and Cultural Organization's (UNESCO's) International Declaration on Human Genetic Data (2003)* and the *Organisation for Economic Co-operation and Development's Principles and Guidelines for Access to Research Data from Public Funding (2007)* [26]. Still, the mentioned establishments and their published rules are unreliable and incomplete because not one of these regulations has a "supranational" position, power or assertiveness.

References

1. Mascalzoni, D., Dove, E. S., Rubinstein, Y., Dawkins, H. J., Kole, A., McCormack, P., Woods, S., Riess, O., Schaefer, F., Lochmüller, H., Knoppers, B. M., & Hansson, M. (2016). International charter of principles for sharing biospecimens and data. *European Journal of Human Genetics, 24*(7), 1096. https://doi.org/10.1038/ejhg.2015.237
2. De Souza, Y. G., & Greenspan, J. S. (2014). Biobanking past, present and future: Responsibilities and benefits. *AIDS, 27*(3), 303–312. https://doi.org/10.1097/QAD.0b013e32835c1244
3. Watson, P. H., Nussbeck, S. Y., Cater, C., O'Donoghue, S., Cheah, S., Matzke, L. A. M., Barnes, R. O., Bartlett, J., Carpenter, J., Grizzle, W. E., Johnston, R. N., Mes-Masson, A.-M., Murphy, L., Sexton, K., Shepherd, L., Simeon-Dubach, D., Zeps, N., & Schacter, B. (2014). A framework for biobank sustainability. *Biopreservation and Biobanking, 12*(1), 60–68. https://doi.org/10.1089/bio.2013.0064
4. Cormier, C. Y., Mohr, S. E., Zuo, D., Hu, Y., Rolfs, A., Kramer, J., Taycher, E., Kelley, F., Fiacco, M., Turnbull, G., & LaBaer, J. (2010). Protein structure initiative material repository: An open shared public resource of structural genomics plasmids for the biological community. *Nucleic Acids Research, 38*, 743–749. https://doi.org/10.1093/nar/gkp999
5. Cormier, C. Y., Park, J. G., Fiacco, M., Steel, J., Hunter, P., Kramer, J., Singla, R., & LaBaer, J. (2011). PSI: Biology-materials repository: A biologist's resource for protein expression plasmids. *Journal of Structural and Functional Genomics, 12*(2), 55–62. https://doi.org/10.1007/s10969-011-9100-8
6. Seiler, C. Y., Park, J. G., Sharma, A., Hunter, P., Surapaneni, P., Sedillo, C., Field, J., Algar, R., Price, A., Steel, J., Throop, A., Fiacco, M., & LaBaer, J. (2014). DNASU plasmid and PSI: Biology-materials repositories: Resources to accelerate biological research. *Nucleic Acids Research, 42*, D1253–D1260. https://doi.org/10.1093/nar/gkt1060
7. Lamesch, P., Li, N., Milstein, S., Fan, C., Hao, T., Szabo, G., Hu, Z., Venkatesan, K., Bethel, G., Martin, P., Rogers, J., Lawlor, S., McLaren, S., Dricot, A., Borick, H., Cusick, M. E., Vandenhaute, J., Dunham, I., Hill, D. E., & Vidal, M. (2007). hORFeome v3.1: A resource of human open reading frames representing over 10,000 human genes. *Genomics, 89*(3), 307–315. https://doi.org/10.1016/j.ygeno.2006.11.012
8. Tapia-Conyer, R., Kuri-Morales, P., Alegré-Díaz, J., Whitlock, G., Emberson, J., Clark, S., Peto, R., & Collins, R. (2006). Cohort profile: The Mexico City prospective study. *International Journal of Epidemiology, 35*(2), 243–249. https://doi.org/10.1093/ije/dyl042
9. Moe, J. L., Pappas, G., & Murray, A. (2017). Transformational leadership, transnational culture and political competence in globalizing health care services: A case study of Jordan's king Hussein cancer center. *Globalization and Health, 3*(11). https://doi.org/10.1186/1744-8603-3-11
10. H3Africa. (2021). Retrieved March 12, 2021, from https://h3africa.org
11. Hewitt, R. E. (2011). Biobanking the foundation of personalized medicine. *Current Opinion in Oncology, 23*(1), 112–119. https://doi.org/10.1097/CCO.0b013e32834161b8
12. CNIO Stop Cancer (2021). *The CNIO biobank.* Retrieved March 12, 2021, from http://www.cnio.es/ing/grupos/plantillas/presentacion.asp?grupo=50004308
13. Wales Cancer Bank. (2021). Retrieved March 12, 2021, from http://www.walescancerbank.com
14. Gaffney, E. F., Madden, D., & Thomas, G. A. (2011). The human side of cancer biobanking. *Molecular Profiling, 2012*(823), 59–77. https://doi.org/10.1007/978-1-60327-216-2_5

15. Bernini, P., Bertini, I., Luchinat, C., Nincheri, P., Staderini, S., & Turano, P. (2011). Standard operating procedures for pre-analytical handling of blood and urine for metabolic studies and biobanks. *Journal of Biomolecular NMR, 49*(2–3), 231–243. https://doi.org/10.1007/s10858-011-9489-1

16. Yu, X., Bian, X., Throop, A., Song, L., Del Moral, L., Park, J., Seiler, C., Fiacco, M., Steel, J., Hunter, P., Saul, J., Wang, J., Qiu, J., Pipas, J. M., & LaBaer, J. (2014). Exploration of Panviral proteome: High-throughput cloning and functional implications in virus-host interactions. *Theranostics, 4*(8), 808–822. https://doi.org/10.7150/thno.8255

17. Fortin, S., Pathmasiri, S., Grintuch, R., & Deschênes, M. (2011). 'Access arrangements' for biobanks: A fine line between facilitating and hindering collaboration. *Public Health Genomics, 14*(2), 104–114. https://doi.org/10.1159/000309852

18. Guerin, J. S., Murray, D. W., McGrath, M. M., Yuille, M. A., McPartlin, J. M., & Doran, P. P. (2010). Molecular medicine Ireland guidelines for standardized biobanking. *Biopreservation and Biobanking, 8*(1), 3–63. https://doi.org/10.1089/bio.2010.8101

19. Chandrasekar, A., Warwick, R. M., & Clarkson, A. (2011). Exclusion of deceased donors post-procurement of tissues. *Cell and Tissue Banking, 12*(3), 191–198. https://doi.org/10.1007/s10561-010-9184-6

20. Bevilacqua, G., Bosman, F., Dassesse, T., Höfler, H., Janin, A., Langer, R., Larsimont, D., Morente, M. M., Riedman, P., Schirmacher, P., Stanta, G., Zatloukal, K., Caboux, E., & Hainaut, P. (2010). The role of the pathologist in tissue banking: European consensus expert group report. *Virchows Archiv, 456*(4), 449–454. https://doi.org/10.1007/s00428-010-0887-7

21. Campbell, B., Thomson, H., Slater, J., Coward, C., Wyatt, K., & Sweeney, K. (2007). Extracting information from hospital records: What patients think about consent. *Quality and Safety in Health Care, 16*(6), 404–408. https://doi.org/10.1136/qshc.2006.020313

22. Garrison, N. A., & Cho, M. K. (2013). Awareness and acceptable practices: IRB and research reflections on the Havasupai Lawsuit. *AJOB Primary Research, 4*(4), 53–63. https://doi.org/10.1080/21507716.2013.770104

23. Colledge, F., Elger, B., & Howard, H. C. (2013). A review of the barriers to sharing in biobanking. *Biopreservation and Biobanking, 11*(6), 339–346. https://doi.org/10.1089/bio.2013.0039

24. Colledge, F., Persson, K., Elger, B., & Shaw, D. (2013). Sample and data sharing barriers in biobanking: Consent, committees, and compromises. *Annals of Diagnostic Pathology, 18*(2), 78–81. https://doi.org/10.1016/j.anndiagpath.2013.12.002

25. Clément, B., Yuille, M., Zatloukal, K., Wichmann, H. E., Anton, G., Parodi, B., Kozera, L., Bréchot, C., Hofman, P., & Dagher, G. (2014). EU-US Expert Group on cost recovery in biobanks. *Science Translational Medicine, 6*(261), 261fs45. https://doi.org/10.1126/scitranslmed.3010444

26. UNESCO (2019). Retrieved March 12, 2021, from http://portal.unesco.org/en/

Methods of Implementation and Set-Up of National Biobanking Networks

Svetlana Gramatiuk, Berthold Huppertz, Mykola Alekseenko, Gabriele Hartl, Tanja Macheiner, Tamara Sarkisian, Zisis Kozlakidis, Erik Steinfelder, Christine Mitchell, and Karine Sargsyan

Abstract

A multidisciplinary retrospective survey of the biobank market with managers and directors or individual researchers in Ukrainian biobanks was launched to collect information on economic models for biosafety. The survey focused on topics such as biobank demographics and structure, sample quality control, funding sources and costing. Consequently, interviews to refine the gathered information and increase the possible market covered were started. The focus of the dialogue lied in the communication between biobanks in order to establish

An example from the Ukraine

S. Gramatiuk (✉) · M. Alekseenko
Institute of Cellular Biorehabilitation, Kharkiv, Ukraine

B. Huppertz
Division of Cell Biology, Histology and Embryology, Gottfried Schatz Research Center, Medical University of Graz, Graz, Austria

G. Hartl · T. Macheiner · C. Mitchell
International Biobanking and Education, Medical University of Graz, Graz, Austria

T. Sarkisian
Department of Medical Genetics, Yerevan State Medical University, Yerevan, Armenia

Z. Kozlakidis
International Agency for Research on Cancer, World Health Organization, Lyon, France

E. Steinfelder
Thermo Fisher Scientific, Waltham, MA, USA

K. Sargsyan
International Biobanking and Education, Medical University of Graz, Graz, Austria

Department of Medical Genetics, Yerevan State Medical University, Yerevan, Armenia

Ministry of Health of the Republic of Armenia, Yerevan, Armenia

© The Author(s), under exclusive license to Springer Nature Switzerland AG 2022
K. Sargsyan et al. (eds.), *Biobanks in Low- and Middle-Income Countries: Relevance, Setup and Management*, https://doi.org/10.1007/978-3-030-87637-1_7

a biobank network. Thus, this chapter shows the process of the establishment of the Ukraine Association of Biobanks with the objective to encourage and develop a Ukrainian biobanking centre to support patients, researchers and industries. The process of building this Biobank Centre was started in 2013 with a bilateral collaboration of only two institutes of medical sciences with the prospect of creating a platform for the association. Currently, it consists of 23 members from central and leading institutes of medical sciences in the Ukraine.

Keywords

Biobank network · Research organization · Market analysis · UAB · SMART IT · SWOT analysis · ASK-Health Biobank · Sustainability plan

A multidisciplinary retrospective investigation of the biobanking market was managed to clarify and collect information on economic models for biosafety. Questionnaires and structured interviews were used for the retrospective investigation in business plans and activities of biobanks. In this way profiles, focuses, collection strategies, finical key aspects, risks and adjoining subjects were analysed.

Questionnaire: Survey Development and Content and Participant Selection

The data on pricing in various biobanks of Europe and the USA were specified on the basis of online pools through e-mail distribution (supported by the Biobank Economic Modeling Tool's (BEMT) questionnaire plan; see Annex), taking into consideration the literature descriptions of the economic characteristics of developing countries [1]. The survey with its detailed questions was divided into five sections:

- Biobank demographics and structure
- Sources of financing and the calculation of costs
- The mechanisms for the formation of prices for samples and data
- The services provided by biobanks
- Quality control samples

The participants of the survey were carefully selected. We created a list of about 90 managers, directors and individual researchers active in biobanking with consideration to ensure a wide geographical coverage. The focus was on the active sides for the most accurate evaluation of the survey data.

The participants in the survey were directors and managers of biobanks that had different and sometimes mixed sources of funding and financing: these included academic medical monoprofile centres, multidisciplinary medical centres, independent academic research organizations and private commercial organizations (pharmaceutical and biotechnology industries). The participants of the survey were experienced in working in biobanks for at least 3 years or more. The contact information of the survey participant was double-checked on the pages of social network (LinkedIn), as well as on other available websites.

The request to participate in the survey was forwarded by a blind e-mail. No personal data of other participants was available to others at this stage. The e-mail comprised a short invitation letter and an information about the survey in general and containing the link to the survey. Responders had the opportunity to save and exit the survey several times and had a time limit of 4–6 weeks to complete the survey submission. Three further e-mail reminders were sent to all participants at 2-week intervals.

Data on general management and possible implementation in biobanks were requested. The following data sources for the identification of the survey answers were applicable: the data from management and governance documents, team meeting protocols (with different responsible staff), strategic concepts, defined activities, the data on quality management system and a strategy matrix.

After collecting the survey replies, interviews started to precise the information and marge the possible covered market.

Structured Interviews for Market Analysis

Biobankers were interviewed regarding profile scope, work areas, collection and storage strategies of human biological material and data. Based on the interviews, we decided to gain more structured information on planning and strategic development methods (if existing), including all typical and well-known aspects of biobanking. Furthermore, biobanking-specific questions have been addressed, to collect concrete information and to be explicitly applicable for the topic of 'building a biobank network-like structure'.

As an orienteer, we used the business plan chapters suggested from Sargsyan et al. as a central document for the interviews, as follows [2, 3]:

1. *Biobank name, including a profile*
2. *Business environment, including the international environment and biobank research projects*
3. *Services of the biobank: service portfolio*
4. *Market, including market segmentation/market analysis and customer groups*
5. *Scientific and economic structures around the biobank*
6. *Development plan of the biobank, including a vision statement, critical success factors and strategic goals such as promotion of relevant study cohorts; cooperation; management; infrastructure and personnel development; ethical, legal and societal issues (ELSI)-dissemination development; and public relations*
7. *Risks and hazards*
8. *Exit strategy and termination possibility, including a cost calculation*
9. *SWOT analysis*

The topics consisted of nine items and were used for investigating the important information on the participating biobanks. These were important strengths, key opportunities, professional levels of employees (if any), exclusive benefits (development of scientific projects, participation in clinical trials), activities or services that the biobank cannot afford, if any, then information about direct competitors, potential threats to the organization of the biobank, etc.

Team Meetings: Development and Implementation

During the revelation of the plan and the analysis of the surveys and the interviews, we realized that the main attention needed to be directed to the dialogue and communication (cooperation) between biobanks, in order to be able to implement an official biobanking network or a comparable construction. Therefore, we organized core team meetings with all leading people of the 23 biobanking initiatives in the Ukraine, addressing the issues listed below:

1. Actions that must be carried out in advance for the safe collection and storage of samples, while long-term prospects and plans are developed
2. Analysis of the state of sample stock and collection possibilities
3. Participation in research programmes for quality control of bio-samples, and if existing utilization of samples of untested quality control
4. Practical methods for sample identification, the possibilities of unification and advancement
5. National plans for information on the work of the biobank—holding a conference, creating a scientific journal on biobanking and consulting the interested medical, pharmaceutical and scientific groups; involvement of experts from Europe for consultation and audits; use of international experience in storage and use of samples; and attraction of experts from international biobank partners
6. Ethical control and development of contracts for the transfer of samples
7. Cooperation with biobanks from countries with low economic development for rare diseases

We had 16 team meetings to get a consensus by not only understanding the past of each partner but also creating a path of joint future. One of the most important findings in the survey and decisions at team meetings was that in the first phase, there was the need to concentrate on the qualitative and not quantitative issues. In this manner, we did not analyse the data on sample numbers or on profiles but rather built up new frameworks and new biobanking environments to work together.

After the creation of the road plan, consultations were held with international partners to determine the most appropriate financing model and to search for the necessary resources. The meeting was held in Vienna, Austria, at the time of the European Biobank Week (EBW) in 2016 during the framework for the most appropriate financing model and the search of the necessary resources. The attendees came from biobanks (Biobank Graz, Integrated Biobank of Luxembourg, Brain

Cancer Biobank, Kazakhstan, National BioService, the first Russian Biobank), clinical organizations, biological resource centres, industrial partners and advocacy groups, including 53 participants from LMICs.

In the second phase of the project, a number of relevant subregional biobanking initiatives were raised. We decided that the biobanking mechanisms in the Ukraine should not set limits on the number of samples that can be stored. We developed a five-step model for implementing networks. Furthermore, we decided to implement a Ukraine Association of Biobanks (the most suitable form that has also the legacy to act in the Ukraine).

The Example of the Ukraine Association of Biobanks

Development and Progress of Ukraine's Biobank Network

The Ukraine Association of Biobanks (UAB) was established in 2017 to encourage and progress a Ukrainian biobank complex to support patients, researchers and industry. In 2013, the Biobank Centre started a bilateral collaboration of only two institutes of medical sciences with the prospect of creating a platform for the association.

Currently, the network consists of 23 members from the central and leading institutes of medical sciences in the Ukraine. The Ukraine Association of Biobanks organized the important processes, built on the ESBB, ISBER and National Cancer Institute (NCI) guidelines, which were harmonized and standardized by the member biobanks. Different guidelines, policies, regulations and documents were developed (such as a patient consent guiding standard, a patient information form, a biobank consent document, an access application form and others) and mutually agreed on (after robust discussion among partners of the UAB) to be used by each partner organization. Also important was the outlining of an optimum categorization for a document submission for the review of the Steering Committee of the Ukraine Association of Biobanks (SCUAB).

The objectives of the Ukraine Association of Biobanks were stated as follows:

1. To develop a Ukrainian biobanking network in the form of an association while considering international guidelines and expertise,
2. To offer fair access to consistently and homogenously obtain. process, and store human biological material and corresponding data for researchers working independently in the academic world or the industry. Important was the involvement of UAB partners in national and international research activities (but only if they are scientifically and ethically approved).
3. To develop technologies, which may lead to better and more accurate diagnostic processes and improved innovative-targeted therapies, but still in parallel stimulate Ukraine's economy, through a genuine collaboration among members, partners and users.

4. To install a biobank network in the form of an association to govern, develop and sustain the UAB as part of the patient care corridor in Ukraine's principal hospitals.

Biobank Network Set-Up

During the last 6 years, consuming the structure of European and American biorepositories as prototypes, the installation of autonomously functioning biobanks happened almost at the same time in large Ukrainian educational hospitals (see Fig. 7.1).

The procedure of establishing an association of biobanks in the Ukraine was developed gradually, persistent mostly with efforts of one to two protagonists, jointly working for mutual goals also on the personal side. Inadequate financing as well as the lack of authorized structure based on law led to a very weak development.

Once a consensus had been achieved between the partners, the Ukraine Association of Biobanks communicated with the responsible ethics committees as well as the data protection authorities, patient representatives, researchers and other stakeholders, to announce cooperation willingness and openness.

The topics that were discussed by the UAB comprised the implementation of probable standards of informed consent in the whole network as well as a possible return of noteworthy investigation results to donors. The Ukraine Association of Biobanks is moulded using examples of international networks such as Marble Arch International Working Group, ISBER, ESBB, etc. The Ukraine government remains careful by supporting, funding as well as directly implementing national biobanking.

The UAB presently consists of 23 members and covers all 7 leading Ukrainian cancer hospitals and several university hospitals. That is why the UAB network emphases are primarily on cancer but also on metabolic sickness samples. From over 25,000 patients, several samples and data have already been collected and are open for research.

The subsequently listed types of cancers were collected at the UAB and are available for research purposes: adrenal gland, bladder, endometrium, kidney, lung, melanoma and non-melanomatous skin cancers, oesophagus, ovary, prostate, testis and thyroid.

The editor, Dr. Svetlana Gramatiuk, ASK-Health Biobank Co-founder, was appointed as the Head of Biobanking and Biospecimen Sciences group at the UAB and was elected as the president of the UAB.

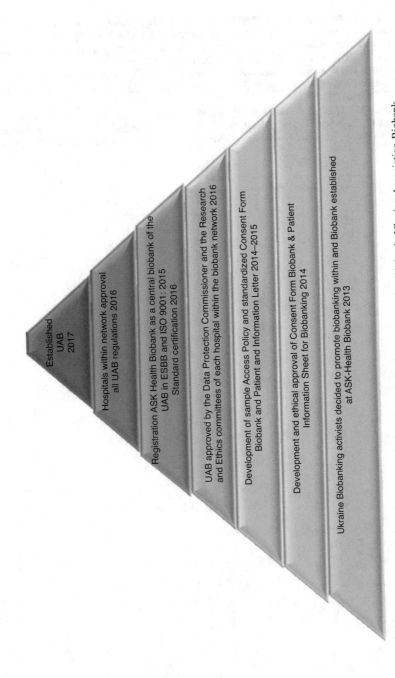

Established
UAB
2017

Hospitals within network approval
all UAB regulations 2016

Registration ASK Health Biobank as a central biobank of the
UAB in ESBB and ISO 9001: 2015
Standard certification 2016

UAB approved by the Data Protection Commissioner and the Research
and Ethics committees of each hospital within the biobank network 2016

Development of sample Access Policy and standardized Consent Form
Biobank and Patient and Information Letter 2014–2015

Development and ethical approval of Consent Form Biobank & Patient
Information Sheet for Biobanking 2014

Ukraine Biobanking activists decided to promote biobanking within and Biobank established
at ASK-Health Biobank 2013

Fig. 7.1 Steps in development of the Ukraine Biobank Network and Ukraine Association Biobank. Ukraine Association Biobank

References

1. Odeh, H., Miranda, L., Rao, A., Vaught, J., Greenman, H., McLean, J., Reed, D., Memon, S., Fombonne, B., Guan, P., & Moore, H. M. (2015). The biobank economic modeling tool (BEMT): Online financial planning to facilitate biobank sustainability. *Biopreservation and Biobanking, 16*(6), 421–429. https://doi.org/10.1089/bio.2015.0089
2. Macheiner, T., Huppertz, B., Bayer, M., & Sargsyan, K. (2017). Challenges and driving forces for business plans in biobanking. *Biopreservation and Biobanking, 15*(2), 121–125. https://doi. org/10.1089/bio.2017.0018
3. Sargsyan, K., Macheiner, T., Story, P., Strahlhofer-Augsten, M., Plattner, K., Riegler, S., Granitz, G., Bayer, M., & Huppertz, B. (2015). Sustainability in biobanking: Model of biobank Graz. *Biopreservation and Biobanking, 13*(6), 410–420. https://doi.org/10.1089/bio.2015.0087

Implementation of Ethical and Legal Considerations in a Biobanking Network

Svetlana Gramatiuk, Mykola Alekseenko, Tanja Macheiner, Christine Mitchell, and Karine Sargsyan

Abstract

The first step in establishing a solid biobanking recourse is clearly to build a solid ethical and legal basis for the operation of each biobank and for a biobank association as a whole. In this chapter we discuss some general considerations on the topic, give some directions and how to and then show an installation process based on UAB. Over the last 5 years, the National Scientific Centre for Medical and Biotechnical Research was the main organization at the Commission on Bioethics at the Cabinet of Ministers of Ukraine that encouraged the formation of authorized requirements for the Ukrainian researchers to work with worldwide agreements with pharma and bioethics. One of the main prerequisites was the coordinated activity of the ethical commission at different levels in the Ukraine. To develop the intended unified national network of biobanks in the Ukraine, the UAB working group was required to submit several separate ethics applications as prerequisite for each participating hospital site and each member organization.

S. Gramatiuk · M. Alekseenko
Institute of Cellular Biorehabilitation, Kharkiv, Ukraine

T. Macheiner · C. Mitchell
International Biobanking and Education, Medical University of Graz, Graz, Austria

K. Sargsyan (✉)
International Biobanking and Education, Medical University of Graz, Graz, Austria

Department of Medical Genetics, Yerevan State Medical University, Yerevan, Armenia

Ministry of Health of the Republic of Armenia, Yerevan, Armenia
e-mail: karine.sargsyan@medunigraz.at

Keywords

Ethics · Ethical approval · Ukraine Association of Biobanks · Bioethics · Legal issues · Local ethical committee · Biomedical research · Standard ethics application form · Scientific and ethical review board

The tendency to store biological material with associated data in biobanks has increased worries concerning how to safeguard the proper moral use of human biological material and the correlated data [1], also to extra broad-spectrum socio-political questions, as the awareness and the acknowledgement of biobanks at large [2]. Certainly, human biobanks have received substantial attention in the ELSI literature during the last decade [3, 4]. Research in ELSI of biobanking has concentrated on questions of confidentiality, informed consent, ownership of human biological material and corresponding data, as well as on benefits of sharing and governance. This is because biobanking-related research challenges the traditional outline for biomedical research on human biological material and its usual components [5]. In prior to deliberating these topics, we explore the doctrines that enlighten ethical discussions in human biomaterial collection and long-time storage. The general principles and values of autonomy and justice in this area of biobanking research are converted into concrete actions by realizing rules and guidelines and setting actions on informed consent, protection of confidentiality and privacy, as well as on non-discrimination procedures, which commonly emerge in the scientific publication and works on ethics of biobanking sciences [6].

The implementation of informed consents into practical working documents faces complications in the matter of broad biobanking activities, especially long-standing storage and inclusion of biological material and corresponding information in research [6]. Even though the donation of human biological material for a particular scientific cause or project collection does not actually post any substantial complication to obtaining ethically approved informed consent, the elaboration of biorepositories and the numerous probable exploitations/utilizations of human materials increase the problematic demand of "how to obtain consent for a multitude of possible research purposes" not known at the time of sample collection [7]. Research participants in that situation cannot be conversant, for example, of the impending dangers and especially advantages of the sample-based science, as the biobanks are usually set for future (jet unknown) aims [8].

Large-scale biobanking must adjust the ethical bases that were developed for project collections and keep the ethical principles [6]. As a result of the high amount of evidence facilitated by biobanking and associated sciences, recognized conceptions of research on medical ethics have been reconsidered, and consent for research, privacy and confidentiality is being restudied [9]. The Declaration of Helsinki mentions "envisioning a specific, discrete research project, not a tool for use on unforeseen and unforeseeable research projects" [6, 10]. This has solidified data protection, confidentiality and privacy main problems in the perspective of research on humans in the twenty-first century and led many researchers (not only ethicists) to think over [11]. If the potential of the biobanking sciences allows sample use for undefined research, linking of biological samples to personal data needs to be restricted, and coding and anonymization is the selected preference for the documentation of human biological material [12]. Several guidelines of the usage of biorepository samples such as an acknowledged permission (informed consent) to all possible potential uses of the material and/or data make the right of persons to choose if and how their samples and related information will be involved in scientific investigations [6]. Progressively, cohesion and exchange—more than autonomy—have appeared as ethical values directorial for informed consent [13]. Expansions in sizable biobanking activities emphasize that the promise of total privacy and discretion is not an assurance that biomedical scientists can provide [14]. This has implications not only for confidentiality but also for privacy [11].

The safety and security of (personal) data is one of the most important actions implemented by biobanks to avoid probable exploit of donor information by employers and insurers [6].

Discussions on ownership of human biological specimens are ongoing, for example, a tissue or a fluid sample in some esteem belongs to the person from who it was taken, as a minimum for an outlined period of time [15]. However, the law relating to ownership and control of human biological material varies in different countries and remains unclear in some countries [9]. Persons may nevertheless keep a specific level of control over their collected samples, mainly as the right to withdraw or to demand obliteration of own samples [10]. Additionally, the participation of an IRB (Institutional Review Board) or REC (Regional Ethics Committee) and the necessity for its positive outlook/vote *"is intended to ensure that an intently worded consent is not exceeded, that consent in broader terms is not incorrectly given an even wider interpretation and that exceptional circumstances in which consent may be waived are not unlawfully invoked"* [11].

For the incorporation of biobanks, outstanding organizations such as IRBs are required to be clear, fast and practical in collaborating with each other [6]. Thus, the results of assessments of IRBs (*votum*) can differ between local committees, regional bodies and country ethical authorities and especially their influences of prosecution depending on local jurisdiction [16]. Also, there is presently no guideline, directive or operating procedure for the joint recognition of IRBs or a pan-European research ethics approval.

The European Forum for Good Clinical Practice (EFGCP) has started a process to encourage good clinical practice (GCP) and encourages the tradition of mutual, high-quality standards in all biomedical research all over Europe [12]. Some initiatives towards conception of common policies and possible standard operating procedures for European RECs also *"need to be encouraged and supported for biobanking integration so that IRBs and RECs can work from a common set of standards, procedures and documentation that will streamline ethical applications without compromising the underlying ethics"* [11].

Data protection authorities likewise have a significant part in supervision of data processing, as in biobanks also for the usage of information and biological materials by researchers [16]. These authorities have been permitted to recognize and classify the significant policies for biobank research in Italy, stipulated direction to biobanking community and stakeholders in Germany and were in charge for observing the creation and process of the Health Sector Database in Iceland [6, 13].

In-house counselling and decision groups have been created in numerous biobanks, which characteristically comprise a scientific advisory board and sample/data access committee [6]. These organizational structures and committees are vital for transparency and liability and assuring self-confidence in the governance of the biobank. As an example, take a look at the UK Biobank board structure, which has situated an autonomous commission, the *Ethics and Governance Council*, as a custodian of the *Ethics and Governance Framework (EGF)*. Under the scope of EGF, the *UK Biobank* gets advice with respect to possible deviations in law and the protection of the rights of biobank donors and the general public [14]. Another example is the Spanish regulation that obligates biobanks to comprise at least two boards of external experts, specifically a scientific advisory board and an ethical board [17]. The expected contribution of these boards is also the information on the ethics and research possibilities and sides of the integration of obtainable biomaterial cohorts within biobanks and the relocation of material to another biobank or a user research institution. The simplification of transparency is further increased by publication of the named persons as affiliate of the external committees to raise public trust.

Public trust is one of the critical and crucial factors for determining whether donors of human biological material are going to support biobank research [18]. Decreased sureness in biobanking practice may have detrimental concerns: *"If individuals start withdraw their consents the human biobanks will not be complete, the possibility to draw scientifically valid conclusions will decrease, and the potential for follow-up examinations and medical treatment will not be fulfilled"* [19]. The progression of construction of confidence in biobanking in general is the key to all kinds of biobanking initiatives in different countries, regardless of whether they trust on patients or the broad public [20].

In many countries, the trust in biobanks is still instable. A statement on conclusions of a *2010 Eurobarometer* survey on the *Life Sciences and Biotechnology* provided information about public perception of biobanks in European countries [21]. Amazingly, the research findings showed only little consciousness on biobanks of majority of Europeans. Over 60% of surveyed persons had no knowledge about biobanks at all, and only 17% were pronounced as having active knowledge in biobanking, collected during debates, consultations or requesting information about biobanking activities. The most knowledgeable respondents were found in the Nordic European countries, such as Sweden, Finland and Iceland. Also, the readiness to give broad consent is directly related to the knowledge about biobanks, while the commercial use remains a concern.

The regulatory characteristics for approvals for the distribution of human biological samples are highly complex and are founded on numerous principles, guidelines, protocols and laws [22]. The issue is becoming trickier and further complex if a transnational aspect is present; regulation regarding the donation of human biological samples and probable compensation of donors differ extensively in countries [23].

Commercial partners associated with biobanks and biobanking activities are seen in general as being to a lesser extent reliable, with worries concerning the unbalanced distribution of remunerations, biased scientific goals and potential misappropriation of individual data [24]. The belief that the human body cannot be an item to profitmaking or a source of commercial advantage is preserved in different commanding papers. The furthermost prominent is the *Council of Europe's Convention for the Protection of Human Rights and Dignity of the Human Being with Regard to the Application of Biology and Medicine: Convention on Human Rights and Biomedicine* [25], which is a cornerstone of bioethics and human rights. Article 21 of the Convention ("Prohibition of financial gain") says: "*The human body and its parts shall not, as such, give rise to financial gain*". As addition, the Explanatory Report [25] clarifies the sense of "*body parts, e. g. nails, which are discarded tissues, and the sale of which is not an affront to human dignity*". Blood is hence unambiguously embraced. Other important documents also repeat about human body and its parts, including blood [26] and no financial gain; some announcements by the *United Nations Educational, Scientific and Cultural Organization*, predominantly the *Universal Declaration on the Human Genome and Human Rights* [27] and also the *International Declaration on Human Genetic Data* [28] are reprising the norm of non-commercialization and use prohibition for profit.

Regarding possible financial profit, different regulations refer to the possibility of the patentability of biological samples and also bring in some other considerations. There is a huge gap of understanding and usage, and the concept of expert centres of BBMRI [29] is starting to solve this problem. There are many efforts actually placed on the implementation of these ideas into reality, and the coming years will hopefully show some results.

In order to illustrate the set-up of the ethical requirements with a real example, the following text describes the step-by-step implementation at UAB.

The first step of the progress of the Ukraine Association of Biobanks was clearly on building a solid ethical and legal basis for operations of each single biobank and the whole association.

During the last 5 years, the *National Scientific Centre for Medical* and Biotechnical Research was the main organization at the *Commission on bioethics at the Cabinet of Ministers of Ukraine (Decision of the Cabinet of Ministers of Ukraine from 13.12.01 № 1677)* that encouraged the formation of authorized requirements for the Ukrainian researchers to work with worldwide agreements with pharma and bioethics. One of the major preconditions was the coordinated ethical committee's action at different stages in the Ukraine. Nowadays, the *Bioethics Committee at the Presidium of the National Academy of Science of the Ukraine performs all the mentioned tasks (Decision of the Presidium of NAS of Ukraine from 07.11.2007 № 288)* [30].

The legal issues in the Ukraine are indirectly regulated by the law on transplantation, and the Law of Ukraine "*On Licensing of the Certain Types of Activities*" Article 9, Part 3, paragraph 22, provided that such type of activity is included: "*The activities of banks of umbilical cord blood, other human tissues and cells"* should be licensed. The Decree of the Ministry of Health of 10.04. 2012, № 251, approved *"Licensing Terms for the Activity of Banks of Umbilical Cord Blood, Other Human Tissues and Cells"*. However, legislation delegated some functions such as collecting samples outside of clinical trials to ethical commissions.

Bearing in mind that actually there are no specific laws, guidelines and regulations in the Ukraine to regulatory affect the formation, the conception, the active work and/or the management of biobanks, which are derived mostly from cross-institutional or mono-institutional research projects, it is becoming clear why major ethical concerns were rising at the beginning of the process. These difficult decisions and even the legal matters are often left to be controlled and accomplished by the local ethical committees and their host institutions.

Informed consent forms and patient information flyers with consistent and complete information, written in an understandable language and terminology, were designed, printed and implemented in all participating organizations exceptionally for the permission of a (biological residual materials) specimen collection for biobanking in the Ukraine. The residual materials of human nature are obtained from leftover of pathological diagnosis procedure after surgery (e.g. operative removal and/or diagnostic biopsies of oncological illnesses). As in any other country, patients were asked to donate blood and/or tissue samples for specific research projects (with a given aim and end date) and were informed about certain topics concerning the project(s). The request to the patients about the donation of human biological samples for a prospective, unknown and undefined research intent is representing a huge shift in the public approach to biomedical research in the Ukraine.

To develop the intended unified national network of biobanks in the Ukraine, the UAB working group was required to submit several separate ethic applications as prerequisite for each participating hospital site and each member organization. The contemporary introduction of a standard ethics application form within UAB was an essential improvement, which accelerated the procedure of standardization and harmonization of operation of member biobanks (see Fig. 8.1).

Good scientific practices are the basis for initiating international cooperation, which can enable a faster progress in the Ukrainian biomedical research landscape in the field of biobanking. These practices were also considered and implemented as a part of ethical-moral responsibility commitments of the UAB. In addition, simple coding instructions, which can easily protect the privacy and confidentiality of donors, have been developed and implemented. The documentation on ethical concerns has a common solution for all biorepositories and researchers of the UAB.

Another important decision was that the leadership of the Ukrainian Association of Biobanks decided to implement a scientific and ethical review board (and also a standard application form), which will be crucial in bringing the balance of interest of members related to various medical research protocols.

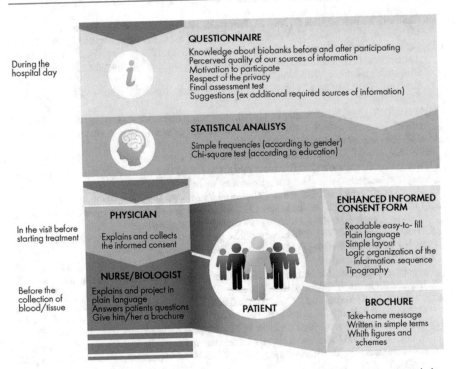

Fig. 8.1 Scheme of how donors for biobanks are informed in the Ukraine. Ukraine Association Biobank

References

1. Cambon-Thomsen, A. (2004). The social and ethical issues of post-genomic human biobanks. *Nature Reviews Genetics, 5*(11), 866–873. https://doi.org/10.1038/nrg1473
2. Gottweis, H., Gaskell, G., & Starkbaum, J. (2011). Connecting the public with biobank research: Reciprocity matters. *Nature Reviews Genetics, 12*(11), 738–739. https://doi.org/10.1038/nrg3083
3. Hawkins, A. K., & O'Doherty, K. C. (2011). "Who owns your poop?" Insight regarding the intersection of human microbiome research and the ELSI aspects of biobanking and related studies. *BMC Medical Genomics, 4*, 72. https://doi.org/10.1186/1755-8794-4-72
4. Tutton, R. (2010). *Biobanking: Social, political and ethical aspects.* Wiley. https://doi.org/10.1002/9780470015902.a0022083
5. Nuffield Council on Bioethics. (2011). *Solidarity: Reflections on an emerging concept in bioethics.* Nuffield Council on Bioethics.
6. Fullerton, S. M., Anderson, N. R., Guzauskas, G., Freeman, D., & Fryer-Edwards, K. (2010). Meeting the governance challenges of next-generation biorepository research. *Science Translation Medicine, 2*(15). https://doi.org/10.1126/scitranslmed.3000361
7. Chadwick, R., & Berg, K. (2001). Solidarity and equity: new ethical frameworks for genetic databases. *Nature Reviews Genetics, 2*(4), 318–321. https://doi.org/10.1038/35066094
8. Mittelstadt, B., & Floridi, L. (2016). *The ethics of biomedical big data.* issn:978-3-319-33523-0.

9. Caulfield, T., Burningham, S., Joly, Y., Master, Z., Shabani, M., Borry, P., Becker, A., Burgess, M., Calder, K., Critchley, C., Edwards, K., Fullerton, S., Gottweis, H., Hyde-Lay, R., Illes, J., Isasi, R., Kato, K., Kaye, J., Knoppers, B., . . . Zawati, M. (2014). A review of the key issues associated with the commercialization of biobanks. *Journal of Law and the Biosciences, 1*(3), 94–110. https://doi.org/10.1093/jlb/lst004

10. Caulfield, T., & McGuire, A. (2013). Policy uncertainty, sequencing, and cell lines. *G3, 3*(8), 1205–1207. https://doi.org/10.1534/g3.113.007435

11. Watson, W., Kay, E., & Smith, D. (2010). Integrating biobanks: Addressing the practical and ethical issues to deliver a valuable tool for cancer research. *Nature Reviews Cancer, 10*, 646–651. https://doi.org/10.1038/nrc2913

12. EFGCP. (2021). Retrieved May 19, 2021, form http://www.efgcp.eu/

13. European Commission. (2021). *Justice and fundamental rights.* Retrieved May 19, 2021, from http://ec.europa.eu/justice/dataprotection/index_en.htm

14. UK Biobank Ethics and Governance Council. http://www.egcukbiobank.org.uk/

15. Lunshof, J., Chadwick, R., & Church, G. (2008). From genetic privacy to open consent. *Nature Reviews Genetics, 9*, 406–411. https://doi.org/10.1038/nrg2360

16. Greely Henry, T. (2007). The uneasy ethical and legal underpinnings of large-scale genomic biobanks. *Annual Review of Genomics and Human Genetics, 8*, 343–364.

17. Arias-Diaz, J., Martin-Arribas, M. C., Garcia del Pozo, J., & Alonso, C. (2013). Spanish regulatory approach for biobanking. *European Journal of Human Genetics, 21*, 708–712. https://doi.org/10.1038/ejhg.2012.249

18. Caulfield, T., McGuire, A. L., Cho, M., Buchanan, J. A., Burgess, M. M., Danilczyk, U., Diaz, C. M., Fryer-Edwards, K., Green, S. K., Hodosh, M. A., Juengst, E. T., Kaye, J., Kedes, L., Knoppers, B. M., Lemmens, T., Meslin, E. M., Murphy, J., Nussbaum, R. L., Otlowski, M., . . . Timmons, M. (2008). Research ethics recommendations for whole-genome research: Consensus statement, 2008. *PLoS Biology, 6*(3), 430–435. https://doi.org/10.1371/journal.pbio. 0060073

19. Hansson, S. O. (2009). Should we protect the most sensitive people? *Journal of Radiological Protection, 29*(2), 211–218. https://doi.org/10.1088/0952-4746/29/2/008

20. Cambon-Thomsen, A., Rial-Sebbag, E., & Knoppers, B. M. (2007). Trends in ethical and legal frameworks for the use of human biobanks. *European Respiratory Journal, 30*(2), 373–382. https://doi.org/10.1183/09031936.00165006

21. Gaskell, G., & Stares, S. (2010). *Europeans and biotechnology in 2010: Winds of change, European Commission, Eurobarometer, EUR 24537.* Publications of the European Union.

22. Soini, S., Aymé, S., & Matthijs, G. (2008). Public and professional policy committee and patenting and licensing committee. Patenting and licensing in genetic testing: Ethical, legal, and social issues. *European Journal of Human Genetics, 16*(1). https://doi.org/10.1038/ejhg. 2008.37

23. Farrugia, A., Penrod, J., & Bult, J. M. (2010). Payment, compensation and replacement—The ethics and motivation of blood and plasma donation. *Vox Sanguinis, 99*(3), 202–211. https://doi. org/10.1111/j.1423-0410.2010.01360.x

24. Gaskell, G., Gottweis, H., Starkbaum, J., Gerber, M., Broerse, J., Gottweis, U., Hobbs, A., Helen, I., Paschou, M., Snell, K., & Soulier, A. (2013). Publics and biobanks: Pan-European diversity and the challenge of responsible innovation. *European Journal of Human Genetics, 21*, 14–20. https://doi.org/10.1038/ejhg.2012.104

25. Council of Europe Portal. (1997, April 4). *Convention for the protection of human rights and dignity of the human being with regard to the application of biology and medicine: Convention on human rights and biomedicine.* Council of Europe, Oviedo. Retrieved May 19, 2021, from http://conventions.coe.int/Treaty/en/Treaties/html/164.htm

26. O'Mahony, B., & Turner, A. (2012). The Dublin Consensus Statement 2011 on vital issues relating to the collection and provision of blood components and plasma-derived medicinal products. *Vox Sanguinis, 102*(2), 140–143. https://doi.org/10.1111/j.1423-0410.2011.01528.x

27. United Nations Educational, Scientific and Cultural Organization (UNESCO). (1997). *Universal declaration on the human genome and human rights.* Retrieved June 1, 2012, from http://www.unesco.org/new/en/social-and-humansciences/themes/bioethics/human-genome-and-human-rights/
28. United Nations Educational, Scientific and Cultural Organization (UNESCO). (2003). *International declaration on human genetic data.* Retrieved June 1, 2012, from http://www.unesco.org/new/en/social-and-humansciences/themes/bioethics/human-genetic-data/
29. BBMRI. (2021). Retrieved May 19, 2021, from www.bbmri.eu
30. National Academy of Science of Ukraine. (2017). *Scientific center for medical and biotechnical research.* Retrieved July 22, 2018, from http://biomed.nas.gov.ua

Collection and Management of Samples

9

Karine Sargsyan, Svetlana Gramatiuk, Mykola Alekseenko,
Tanja Macheiner, Gabriele Hartl, Tamara Sarkisian, Zisis Kozlakidis,
Erik Steinfelder, Armen Muradyan, Christine Mitchell,
and Berthold Huppertz

Abstract

A number of sample access policies from member biobanks and also from domestic biological repositories were considered during the development of the UAB Model Tax Receipt Policy. Developing a SAP was well-thought-out and required delicate negotiations within the UAB. Human genetics research is an

Access Policy in the Ukrainian Biobank Network

K. Sargsyan (✉)
International Biobanking and Education, Medical University of Graz, Graz, Austria

Department of Medical Genetics, Yerevan State Medical University, Yerevan, Armenia

Ministry of Health of the Republic of Armenia, Yerevan, Armenia
e-mail: karine.sargsyan@medunigraz.at

S. Gramatiuk · M. Alekseenko
Institute of Cellular Biorehabilitation, Kharkiv, Ukraine

T. Macheiner · G. Hartl · C. Mitchell
International Biobanking and Education, Medical University of Graz, Graz, Austria

T. Sarkisian
Department of Medical Genetics, Yerevan State Medical University, Yerevan, Armenia

Z. Kozlakidis
International Agency for Research on Cancer, World Health Organization, Lyon, France

E. Steinfelder
Thermo Fisher Scientific, Waltham, MA, USA

A. Muradyan
Yerevan State Medical University, Yerevan, Armenia

B. Huppertz
Division of Cell Biology, Histology and Embryology, Gottfried Schatz Research Center, Medical University of Graz, Graz, Austria

important area when it comes to biobanks. Also within the UAB working group, it was clear that one of the most important benefits is the establishment of member biobanks and their specific collection of samples and data. This is important in order to work with modern applications in biomedical research at an international level and to study the range of human genetic variations. SAP systems were reconsidered based on the sample access ordered by the Ukrainian biobanks. An important step within the Ukrainian Biobank Association was the establishment of a multidisciplinary steering committee. Experts from each research group form the SCUAB sample access committee. SCUAB consists of physicians, scientists, and biobanking specialists who can provide information on regional, general, and international biobanking queries.

Keywords

Sample access policies (SAPs) · SCUAB · Management and collection of samples · Sample input · Sample output · Sample distribution · Sample access committee

Several sample access policies (SAPs) from member biobanks of the UAB and also from international biorepositories were reviewed in formulating of the UAB's sample access policy. The development of an SAP has been very deliberate and has required delicate negotiations within the UAB. They detained differentiations of eligibility factors for the access of samples and corresponding data. Also important was that the consensus on the priority for the Ukrainian research groups was time-consuming, corresponding with active discussions and opinion exchange. A comprehensible SAP is a pivotal milestone for a well-functioning association as for any other consortia and must represent and provide a simple and transparent instrument for accessing samples for researchers.

The development of a multidisciplinary Steering Committee of the Ukraine Association of Biobanks (SCUAB) was the next major step. The SCUAB reviews applications for accessing samples and assisted data from biomedical research groups from academia and industry. Representatives from each research group of collecting sites are nominated for the SCUAB.

The SCUAB is a multidisciplinary and multi-professional active team of clinicians, scientists, and specialists in biobanking, which is recognizing local, but also national, and international significance and an impact on biobanking. The most important task of SCUAB is reviewing applications from researchers, not only from academia but also from the industry, also from national and international groups for approval (or rejection) of sample access. Also within the UAB, the conduction of large-scale research studies is based on reviewing of the SCUAB, which necessitates involvement of all partners and recruitment of samples from multiple hospitals within the association.

Nevertheless, it has become obvious that an associated partner biobank which constantly shows a preference for local researchers in their own host organization rather than national or international researchers coming from outside of the given site faces complications for requests of human biological samples from other partner biobanks of the association. For that reason, it was mutually agreed that for all member biobanks of the UAB, it is indeed beneficial to facilitate all kinds of collaboration projects.

As was already reported in the introduction section, research on human genetics is one of the noticeable fields where biobanking is involved; sometimes biobanks carry out the research themselves. Also within the working team of the UAB, it was clear that one of the most significant benefits comes from the establishment of the member biobanks and their specific research aimed collections of human biological samples and related data. The work with modern applications in biomedical research on an international level and to study the extent of human genetic variations is very relevant to the members. These biobanks intended to accomplish a potentially very significant role in biomedical and human genetics research, as long as the epidemiologically relevant pool of samples and management of those samples and corresponding data are standardized and suitable for requirements of internationally recognized investigators.

If at all possible, the human biological material should be collected and stored as best as possible to suit and meet the needs of future developments in human genetics and molecular biology research, which can be foreseen from today's perspective. It would be disastrous, given the situation that after a decade, it would be realized that the human biological material, which was collected and stored at the UAB, is not adequate for research. Consequently, understanding the importance of this decision and intending to develop potentially potent resources, the SCUAB is continuously performing literature research to find hits and trends in biomedical research and recommendations and/or providing detailed guidance to members, before a new major sampling effort and/or profile building specimen collection begins.

As soon as a study under the scope of the UAB is started, it is vital that the banking of samples, and, if applicable, transport to the central biobank in Kharkiv, is carried out in a structured, robust, and reliable manner. For this purpose, the active team of the UAB works with each member site and each collecting unit through an IT system called "SMART" to guarantee that all sites use suitable collection methods, harmonized and agreed for the given specific study. Also important is to monitor that they use the same supplies of consumables and the same consumables in all participating sites and that the processing is made at a safe and timely frame.

Under the scope of the UAB, in all member biobanks for a given collection, the same principles and systems are implemented for sample management: shipping samples, clinical collection of samples, sample registration, research sample management, client correspondence communication, sample distribution, research sample qualification, sample preparation, and sample storage (see Fig. 9.1).

Biobank personnel may prepare biomaterials for transport only based on approved SOPs from vendors of trust of the UAB. This is of primary importance to all UAB biobank sample logistics. For human biological material shipping

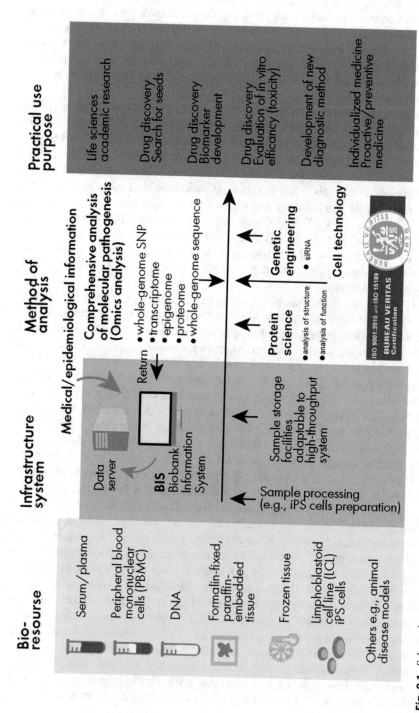

Fig. 9.1 Schematic representation of management and collection at UAB. Ukraine Association Biobank

services, several companies have as a prerequisite the safe and timely transport of biological material within, but also from or to Ukraine from all over the world. To ensure compliance with international standards, various delivery conditions can be used, while vendors and companies support the network to minimize temperature deviations during the transit of samples in the Ukraine, to the Ukraine, and from the Ukraine. The UAB is getting requests for shipment of a variety of materials from the storage site, such as liquid nitrogen, dry ice, wet ice, and samples of temperature storage between 2 and 4 °C to meet the needs of each type of sample.

As already described, the UAB provides detailed SOPs for all procedures in the member biobanks including the collection of samples. The protocols of collection are part of the UAB's shape of sample logistics for all research projects. Standard operating procedures include instructions of specific steps of collection, the tube types, the order in which they are collected and aliquoted, quality control, and the handling and dispatching of mail items. The steps to follow are precisely described (see Annex).

Registration of Samples

The registration of samples at the member biobanks of the UAB is carried out continuously in an IT system called SMART. The SMART system was developed in the Ukraine, taking the economic component and the national and scientific features into account. SMART is creating a unique record at each time when registering samples.

The human biological material is tracked in the system for storage and for receipt regardless of the storage method (tubes, FFPE, fresh frozen, etc.). SMART uses 2D barcodes with a scanner that can identify the information of individual aliquots of the main sample, the donor and collection, even the tube type, and the sample type or tissue, which was entered in the web interface.

Input of Samples

The process of including each sample in the common UAB biobank system is carried out and controlled by the SMART IT system. If the samples are delivered to the central biobank storage facility in Kharkiv, each parcel is scanned upon arrival, and the collection forms and individual samples are checked for accuracy.

During input, samples are automatically assigned to specific projects. All samples have data on the site of collection, the donor (pseudonymized), the clinical and pathology report, and the quantity of material.

Communication of Project Management

The project managers of a given project routinely correspond with sites and researchers to ensure all needs are met and all sample-processing workflows are followed. The clinical and pathology reports are generated in a given time frame. Relationships are built with each research group/client to ensure the projects' needs are being satisfied: parameters of inclusion and exclusion, single parameters of clinical information, the volume of aliquots, the amount of material required, etc.

Distribution

Human biological material provided from members of the UAB has been disseminated worldwide. Members of the UAB are obtaining broad education in international guidelines including International Air Transport Association (IATA)/ shipping regulations. Due to different courier contacts, the UAB is able to ship at any required condition: room temperature, dry ice, etc.

Sample Qualification and Processing

All samples that enter the biobanks are checked for compliance with volume and quality. Samples that do not pass quality control and do not meet the criteria of scientific projects receive a code for rejecting the registration before processing the sample and transferring it to the repository.

At first, the samples arrive in the laboratory department of the biobank; they are allocated to the relevant material department of the laboratory. The lab technician of the biobank processes the material according to the validated standard protocols in accordance with the type of specimen. The processing protocol differs in relation to the type of sample: blood and tissue samples may be used for the isolation of subsamples: buffy coat, mononuclear cell, specific immune cell, and DNA/RNA.

Storage

The storage of samples in accordance with research project-specific conditions of the UAB member biobanks is performed as long-term storage at liquid nitrogen (-180 °C), -80 °C, -20 °C, 4 °C, and room temperature.

The variety of temperature storage facilities is offered by both individual members and the central storage capacity in Kharkiv. All storage systems are maintained and monitored manually, and some important infrastructure units have automatic alarms.

The SMART IT system continuously monitors the temperature in the system, and its programmed freeze/thaw cycle limits the flow of moisture to maintain sample quality. The UAB member biobanks have a number of semiautomated and fully automated analyzers and aliquot systems for processing blood, tissues, and other biological samples while maintaining the eligibility criteria and research project criteria. Most analytical procedures used in UAB member biobanks involve several SOPs.

The automated system of sample tracking with 2D technologies used at the UAB guarantees the elimination of errors with the labelling and sample identification data. These new states of technological innovations provide data that can be tracked and presented through online processes. Stages of sample preparations include automated or manual sample aliquoting, such as blood fractionation (plasma, serum, etc.), aliquoting biomaterial, and preparing tissue. Manual sampling retrieval may involve simply picking a sample from an automated system. The customized service of a biobank is provided for all sample types, volumes, and concentrations. Clients are offered a system for direct interaction with project managers based on research project needs.

SAPs were reconsidered in the sample access policy, commissioned by the Ukrainian biobanks. The disclosure of an SAP required extensive trading with the UAB. Defining illegibleness for specimen admittance and comprehend agreement on antecedence for the Ukraine study assembly was conclusive. A cleaving SAP contributes researchers with an unmingled and diaphanous movement for the admittance of specimens.

The construction of a multidisciplinary steering committee was an important step within the framework of the Ukraine Association of Biobanks. The SCUAB analyzes applications for samples from medical research groups. Specialists from each research group constitute the sample access committee (SAC) for the SCUAB.

The SCUAB is composed of clinicians, study scientists, and specialists for biobanking who are able to learn about regional, general, and international solicitations of biobanking. The SCUAB analyzes applications, submitted both by the nation and the investigators, providing comprehensive-separate pondering that describe specimens from manifold hospitals within the network.

On the other hand, a hospital site consistently prefers local projects to national or international projects, which may cause difficulties during the process of obtaining samples from other biobanks in the frames of the network. However, there are benefits for each place within the network to support collaborations with the members of the UAB and the network having a priority.

Quality Control and Risk Management in Biobanks

Gabriele Hartl, Tanja Macheiner, Svetlana Gramatiuk, Mykola Alekseenko, Tamara Sarkisian, and Karine Sargsyan

Abstract

Human biological materials stored at the UAB are collected according to the general harmonization commitments of all members and according to common standard procedures. Quality control is applied to 100% of the tissue samples collected, which means that requirements are set for the collection, storage and transport of each sample. This process guarantees the highest quality of samples and data. UAB member biobanks support a precise quality control and assurance strategy and are committed to being open to agility and adapting SOPs to innovative developments and for research needs. This chapter describes the processes and methods of implementing quality control at UAB.

Risk management is an essential part of all operational business activities. However, for organizations providing biobanking, reducing risks is inevitably a critical element for daily business. The irreplaceability of human biological

An Example from the Ukraine

G. Hartl · T. Macheiner
International Biobanking and Education, Medical University of Graz, Graz, Austria

S. Gramatiuk · M. Alekseenko
Institute of Cellular Biorehabilitation, Kharkiv, Ukraine

T. Sarkisian
Department of Medical Genetics, Yerevan State Medical University, Yerevan, Armenia

K. Sargsyan (✉)
International Biobanking and Education, Medical University of Graz, Graz, Austria

Department of Medical Genetics, Yerevan State Medical University, Yerevan, Armenia

Ministry of Health of the Republic of Armenia, Yerevan, Armenia
e-mail: karine.sargsyan@medunigraz.at

material, cell lines, associated information and valuable products (e.g. cell-based medicine and biologically effective pharmaceutical components) requires enormously precise planning, comprising analysis of the full range of risk factors. These factors can be categorized as follows: reputation risks, ethical risks, economic risks, operative risks, laboratory risks, personnel risks, infrastructure risks, computer technology and data management risks, strategy and development risks and natural disasters.

Keywords

Quality control · Quality assurance · Core processes · Support processes · Standard operating procedure (SOP) · Risk management · Critical factors · Risk assessment

Introduction

In the biobanks linked to UAB, analysis of the data obtained from different research groups showed that around 80% of researchers and research units use laboratory data from two or more laboratories, which inevitably leads to the need for additional format and error checking for missing or unacceptable values as well as maintaining consistency and adding annotations. It requires the creation of a group of specialist trained scientific supervisors in the extent of supervision and analysis of laboratory data obtained.

Some errors may be evident when tested by medical analysts or by consistency testing based on analytical laboratory programs, but it is required to authenticate the precision of the data by repeating the workplace analysis by random sampling from the samples submitted.

To regulate the quality of laboratory studies, numerous statistical sampling methods were developed. They are based on an independent analysis of at least two or more laboratories of 15% of samples submitted.

Since all data collections enclose errors, and resources are required to minimize these errors, agreeable standards for specific representations of data in the repository must be determined by participating researchers and biostatistics. These error rates are likely to vary depending on the hypotheses and the statistical government needfulness to derive to them. The success and credibility of the work carried out by several participating sites, researchers and laboratories will confide on the evidence that there are reasonable standards for quality check.

Patient data:

- Clinical and demographic
- Collected using case report forms
- Managed by clinical trial research list

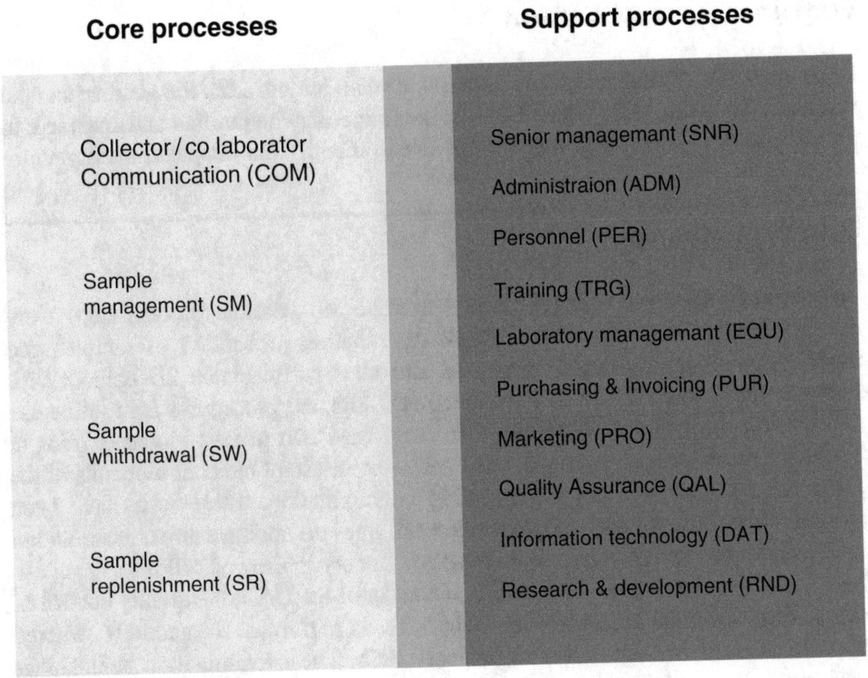

Core processes   **Support processes**

Collector / co laborator   Senior managemant (SNR)
Communication (COM)
                                                Administraion (ADM)

                                                Personnel (PER)

Sample   Training (TRG)
management (SM)
                                                Laboratory managemant (EQU)

                                                Purchasing & Invoicing (PUR)

Sample   Marketing (PRO)
whithdrawal (SW)
                                                Quality Assurance (QAL)

                                                Information technology (DAT)

Sample
replenishment (SR)   Research & development (RND)

Fig. 10.1   Core process and support process. Ukraine Association Biobank

- Standards across all centres
- Data: quality managed and database

The human biological materials that are stored at the UAB are collected according to general commitments of all members for harmonization and according to the common SOPs. Quality control is applied to 100% of the collected tissue samples, which implies the requirements for removing, storing and transporting each sample. This *recambio* mechanism guarantees samples of the highest quality and is delivered to the customers.

The UAB member biobanks are supporting an exact strategy in quality control and warranty procedures while committing agility to be open and to adjust the SOPs for innovative developments and corresponding new needs for the research community. The UAB has implemented highly innovative methods for both divisive and functional quality control and verifying procedures (see Fig. 10.1).

Analytical Quality Control

The analytical quality control performed in member biobanks of the UAB states that the methods and actions are to be deliberated for ensuring lab results and to be analogous, consistent and precise in identified frames.

Functional Quality Control

From each tube during sample preparation, a small aliquot from the genuine sample collection is stored, from which DNA in the future may be purified and analysed, to fix any sample contamination and to be able to identify the sample's identity.

Quality Assurance

To minimize each possible operational mistake, all possible process steps were computerized mostly to exclude cross-contamination probability of samples and incorrect labelling. Collected specimens and vials carry unique 2D barcodes for secure identification. When new reagents, new kits and procedures for routine use are introduced, they are all validated in more than two partner biobanks prior to implementation. All cell lines that exist and were in use in member biobanks of the UAB were tested in-house for detecting contamination. These tests have been successful, and the results have shown the cell lines do not have any contamination (no bacteria, no mycoplasma and no fungus). The cryopreserved cell lines have also been tested on viability. In the member biobanks of the UAB, the quality control of cell lines (contamination as well as viability) was performed in randomly selected vials from each biobank counting at least 10%. The contamination testing was performed by crop growing and polymerase chain reaction (PCR). Using PCR analysis, the DNA derived out of these 10% of cell line samples was amplified with particular primers of microbes.

Sample Management Quality Assurance

Being situated in a developing country, the UAB member biobanks and the central storage facility in Kharkiv are maintaining secure and contain innovative units of redundant backup emergency power generators. Cryopreserved human biological materials, which are of higher interest, are stored in different facilities that contain backup possibilities, with 24/7 monitoring.

Contamination Assessment of Nitrogen Tanks

Longitudinal biobanking of human biological materials in liquid nitrogen can be a source of infection by microorganisms. A group of researchers has established a study for an investigation in microbial contamination of embryo specimens stored for 6–30 years (gas phase of liquid nitrogen). The frazil conglomerate can be a reason of contamination by microorganisms, too. The results of this study have shown that different types of microorganisms (32 bacteria and 1 fungus) could be detected as contamination agents in liquid nitrogen-stored embryos [1]. It is proposed that ice amassing in an opening side of a nitrogen tank can accrue as a catalyser for microbial contaminations [2]. The ice accumulation forms naturally in the atmosphere every time the tank is opened. This ice materialization on cold surfaces such as internal metal constructions or possibly the plastic boxes can be the cause of contamination.

In this context, it was decided to regularly monitor the bacterial and viral contamination of samples from the member biobanks of the UAB. The laboratory departments of the UAB have worked according to Bielanski's process, to check microbial infection of smooth propellant combat tank.

Every 6–8 months, the employees of the bacteriological department take a swipe of smears from containers with liquid nitrogen. Subsequently, bacterial studies are carried out to isolate the pure culture.

Quality Control of Bio-specimens in Biobanks and Hospital Sites

The UAB advised that laboratory departments of hospital sites should develop and execute quality criterions of bio-specimens (fresh-frozen tissues).

Other Activities

Due to numerous actions of the UAB, the control of quality in biological material has been accomplished. For instance, the UAB, its member biobanks and the central storage capacity have applied a quality management system ISO 9001:2015 which is applicable to any organization, regardless of its type or size for the compilation, safety and arrangement of specimens (serum, plasma, urine, buffy coat, etc.), and was certified through the Bureau Veritas. The development and implementation of SOPs for a preservation collection and other activities with bio-samples was carried out for the first time in 2013 to make most SOPs a habit [3]. A storage system, which contains an automated handling unit for organized and efficient storage administration, was obtained in 2016 (300,000 containers at −80 °C temperature) [4]. For the methodical storage management, the machine-controlled bio-specimen management IT platform SMART is used.

The Research of Characteristics

The scientists from research groups within the UAB have been observing different components of the biobanking process influencing the quality of human specimens. This information was incorporated in the best practising SOPs and for a compilation with high-quality human bio-specimens.

The controlling of risks is a segment of all important business processes. Nevertheless, for biobanks and biobanking-performing organizations, risk qualification is inevitably an extremely precarious component for day-to-day procedures. The costly kind of inimitable biobanking materials, corresponding information and other high

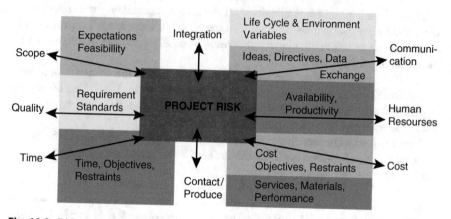

Fig. 10.2 Risk management and biobanking. Ukraine Association Biobank

significance biobanking goods such as cell-based medicines and biologically active pharmaceutical compounds call for enormously precise planning comprising the full spectrum of risks. These risks are usually sectioned into the subsequent classes: ethical risks, reputation risks, economic risks, strategic and developmental risks, human resources (HR) risks, operational risks, typical lab risks, mechanical and machine (infrastructural) risks, IT risks and natural disasters (see Fig. 10.2).

Risk assessments and/or determinations are in principle extremely specific, and they are built on particular awareness of the staff. Therefore, to get a diverse and comprehensive overview about an organization's risks, it is recommended to organize a training with the personnel from all hierarchical levels and all areas of biobanking such as managers, public relations (PR) officers, HR officers, lab technicians, accountants, scientists, IT specialists, bioethics specialists, engineers, medical doctors, research nurses and carrier organization members (as strategy makers and stakeholders).

References

1. Doucet, M., Becker, K. F., Björkman, J., Bonnet, J., Clément, B., Daidone, M.-G., Duyckaerts, C., Erb, G., Haslacher, H., Hofman, P., Huppertz, B., Junot, C., Lundeberg, J., Metspalu, A., Lavitrano, M., Litton, J.-E., Moore, H. M., Morente, M., Naimi, B.-Y., . . . Dagher, G. (2016). Quality matters: 2016 annual conference of the national infrastructures for biobanking. *Bioperservation and Biobanking, 15*(3), 270–276. https://doi.org/10.1089/bio.2016.0053
2. Schenk, M., Huppertz, B., Obermayer-Pietsch, B., Kastelic, D., Hörmann-Kröpfl, M., & Weiss, G. (2016). Biobanking of different body fluids within the frame of IVF—A standard operating procedure to improve reproductive biology research. *Journal of Assisted Reproduction and Genetics, 34*, 383–290. https://doi.org/10.1007/s10815-016-0847-5
3. Bernini, P., Bertini, I., Luchinat, C., Nincheri, P., Staderini, S., & Turano, P. (2011). Standard operating procedures for pre-analytical handling of blood and urine for metabolic studies and biobanks. *Journal of Biomolecular NMR, 49*(2–3), 231–243. https://doi.org/10.1007/s10858-011-9489-1

4. Park, J., Hu, Y., Murthy, T. V. S., Vannberg, F., Shen, B., Rolfs, A., Hutti, J. E., Cantley, L. W., LaBear, J., Harlow, E., & Brizuela, L. (2005). Building a human kinase gene repository: Bioinformatics, molecular cloning, and functional validation. *PNAS, 102*(23), 8114–8119. https://doi.org/10.1073/pnas.0503141102

Governance and Stakeholder Analysis

Svetlana Gramatiuk, Tamara Sarkisian, Zisis Kozlakidis, and Karine Sargsyan

Abstract

For biobanks, stakeholders are medical or pharmaceutical organizations, groups of scientists or individual scientists with interests within the organization or individuals who can influence the organization's operations and policies. Within this chapter, the method of developing a stakeholder analysis for the ASK-Health Biobank from Ukraine is presented. Here, stakeholders were identified and then analysed, assessed and linked to specific influencing factors. Based on these results, interventions and strategies were planned and implemented in a risk management plan.

Keywords

Stakeholder analysis · Identification of stakeholders · Ranking of needs

A Ukrainian Showcase

S. Gramatiuk
Institute of Cellular Biorehabilitation, Kharkiv, Ukraine

T. Sarkisian
Department of Medical Genetics, Yerevan State Medical University, Yerevan, Armenia

Z. Kozlakidis
International Agency for Research on Cancer, World Health Organization, Lyon, France

K. Sargsyan (✉)
International Biobanking and Education, Medical University of Graz, Graz, Austria

Department of Medical Genetics, Yerevan State Medical University, Yerevan, Armenia

Ministry of Health of the Republic of Armenia, Yerevan, Armenia
e-mail: karine.sargsyan@medunigraz.at

Biobanks have interactions on many levels. As biobanks are important long-term research infrastructures that represent the basis for the development of precision medicine (personalized/stratified medicine) in research institutions, the levels of interactions for a biobank are diverse. Globally, they are still perceived as young institutions (although they contain samples from many decades), unknown not only to the general public [1] but also to scientists and clinicians in developing countries, and the sustainability of these institutions depends on support from various communities [2]. Effective management of the interaction with the relevant reference groups based on a stakeholder analysis is therefore an important task in the project management of a biobank [3]. In the scientific literature—especially from the Anglo-American and Northern European regions—the focus on the topic of biobanking communication and participation has so far been on informing and involving patients, test subjects, citizens or "the public". The spectrum of activities and objectives ranges from increasing the awareness of biobanks as a research resource with the aim of legitimizing them socially and improving willingness to participate [4, 5] through the survey of attitudes, perspectives and concerns of sample donors and the public, in order to be able to adapt the argumentation and the offers [5, 6] up to the integration of patients in the steering committee or even the establishment of biobanks by the patients themselves [2, 6]. The importance of other stakeholders—the medical institutions and funding organizations as the main donors or the researchers as the actual "customer" of a biobank [7]—has only recently been given more prominence and systematically researched.

The professional groups active in healthcare who (should) take samples in the treatment context, which are then stored in a central biobank for research purposes, are increasingly coming into focus, because without this kind of biobanks (connected to central labs of multi-profile clinics) and their contribution, the possibilities of sample collection are very limited [8]. Cañada et al. even argue that direct interaction with the public is only a minor part of interaction with stakeholders and that involvement of other groups—hospitals and clinicians, other biobanks, public administrations and sponsors, industry as well as researchers—may be even more important when the biobanks are really intending to generate scientific added value [2].

For communication, for the development phase of centralized biobanks, for example, at university clinics, the following different levels of communication requirements should be taken in account:

(a) Patients and healthy controls as sample donors—need of being informed
(b) Scientists and clinicians—the expectations towards the new infrastructures
(c) Biobank personnel—being informed by managers

The communication and information need and amount should be determined in order to generate the desired commitment. Also, this is why carrying out a stakeholder analysis is a primary task.

Stakeholders are the reference groups or target groups of the organization (or a policy area or topic) who are influenced by it and in turn influence it [3]. Only by identification, characterization and prioritization of these groups is it possible to strategically plan the relationship and interaction with them. Numerous methods of stakeholder analysis have been described, but in principle they proceed in a comparable way [9].

Identifying stakeholders is one of the important first steps in this process of stakeholder management. In practice, this is usually done through a brainstorming session, which should include people who know the thematic environment well [9]. It is strongly advisable to have the primary generated list of stakeholders checked by someone who is not part of the project team (neutral opinion) [3]. Stakeholders of a biobank can be classified as internal or external to the caring organization (university, clinic, company, etc.) or according to the nature of their association with the biobank:

(a) Directly associated (e.g. employees, sample donors)
(b) In the immediate vicinity and with direct interaction (e.g. scientific sample users)
(c) In the wider environment (e.g. other biobanks, authorities, legislators)

Describing each particular stakeholder's interests, values and influence is highly important (e.g. positive/negative attitude), but it is much better and more advisable if they are described on the basis of research data (qualitative, e.g. interviews, quantitative, e.g. surveys, etc.) so that they can be characterized in terms of their possibilities of influencing.

For example, the stakeholders can be mapped in an influence-interest matrix, whereby other decisive organizational features can also be selected for the mapping, if necessary. The stakeholder management and interaction strategies result from the respective placement.

It is important to ensure that the interaction with the people and groups who have a high level of interest and also influence is very close and maximally interactive, while those who have a high level of influence but little interest are concerned with keeping them satisfied. In the case of stakeholders who are very interested but have little influence, it is sufficient to provide them with information on a regular basis (which can be monthly, quarterly, etc. based on the given agreements), while those who are neither interested nor influential should only be observed in order to identify any critical changes at an early stage.

Since resources are usually limited, it is essential to prioritize the stakeholders on the basis of the previous characterization, possibly to summarize different groups, and to limit the strategic planning of measures to selected stakeholder groups.

On this basis, a specific interaction strategy (engagement strategy) must be developed for each of the prioritized stakeholder groups. Hereby, the stakeholder-specific "commitment" can be exactly described and documented in the stakeholder management plan or in a different level model. The spectrum of levels ranges from activities that are intended to change the expectations or views of stakeholders, through educational measures and seeking advice, to partnerships with equal rights and actually shared decision-making power [10, 11]. In today's increasingly partici-patory society, it should always be critically examined what the purpose of the interaction and integration is. Pro forma measures should be avoided.

In order to quickly adapt the measures, if they are not successful or if external conditions change, it is necessary to closely monitor the attitudes of the stakeholders. If the characteristics of the stakeholders change, a change in the interaction strategy or a change in the project plan may be necessary [3].

To present the stakeholder analysis and management plan process in a more applied manner, the UAB stakeholder management process is described. The stakeholders of UAB were identified as medical or pharmaceutical organizations, groups and individual scientists with a concern in the institution, which can affect the organization's actions and policies. The growing number of new biobanks underlines the importance for each and every biobank to identify their stakeholders.

Groups have been built from all participants. They became assigned issues for the development of a hypothetical outline in biobanking, which must include a ranking of needs (e.g. economic, operational or infrastructural). The setting up of a clinical study based on new and comprehensive SOPs was also discussed. The clinical professionals and heads of departments simplified the situation into (for politicians) understandable SWOTs. Results were introduced as follows (Table 11.1).

At the ASK-Health Biobank (Kharkiv), the identification of stakeholders was done in the framework of writing a sustainability plan. Here, stakeholders were identified and subsequently analysed, evaluated and related to specific influencing factors. Based on these results, interventions and strategies were planned and implemented into a risk plan.

The following stakeholders of the ASK-Health Biobank from the Ukraine were identified:

- Stakeholders that are directly associated with the ASK-Health Biobank such as donors and staff of the University's Medical Clinic and Centres of Ukraine Associated Biobank as the owner of ASK-Health Biobank
- Stakeholders within the closer environment (e.g. biobank users, service customers, research partners) were assigned to core activities of the ASK-Health Biobank. These core activities are collecting, processing, storing and providing specimens for information, analysis and preparation, education and training and consulting and coaching.
- Many stakeholders were identified in the wider environment such as other biobanks, legislation, ethics and the general public.

The stakeholder analysis is an important tool for biobanks to identify the optimal future strategy and to evaluate potential risks (see Fig. 11.1). It is important to communicate continuously with stakeholders to be able to act and react to changes quickly. Implementing a stakeholder analysis in the management strategy is recommended to all biobanks.

Table 11.1 Strengths, weaknesses, opportunities and threats analysis

Strengths	Weaknesses
• Strong functional and organized healthcare system	• Finance and support
• Education and exposure of health professionals to rare diseases	• Infrastructure maintenance
• Bio-resources, materials, bio-specimens	• Suboptimal utilization
• Infrastructure buildings	• Coordination of existing infrastructure
• Population literacy—willingness to participate from population/potential donors	• Doctor-to-patient ratio is limiting—heavy clinical load depriving research
• Scientific personnel, expertise and specialists with potential to be trained in biobanking	• Lack of in-country collaborations and networking
• Diversity of pathology	• Lack of core facilities and coordination
• Laws and regulations in place for medical research	• Limited/poor resources for research education
	• Limited possibilities for qualified training of scientists and health professionals
Opportunities	**Threats**
• Health improvement	• Cultural acceptance
• Epidemiology and mechanisms of diseases	• ELSI and trust issues
• Funding from corporate organizations and philanthropists	• Sustainability
• Collaboration opportunities and building capacity	• Ownership of research agenda
• Development of biobank guidelines and framework to ensure interoperability related to laws and policies for medical research that are already in place	• Socioeconomic and political instability
• Streamline the regulations for ethical decisions and approval of MTA for international collaborations	• Long-term control, security, trust, transparency, governance
• Access to samples	• Infrastructure equipment service and maintenance, loss of samples due to power problems, costs, security, transportation logistics
• Technology development	• Biobank supplies and consumables, inconsistencies
	• Adaption of laws and regulations

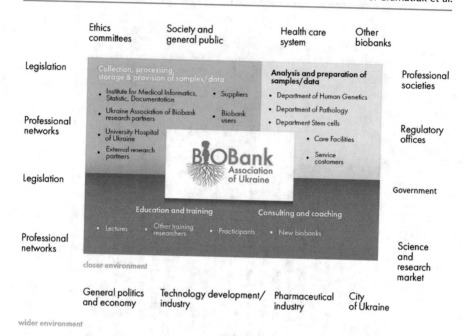

Fig. 11.1 SWOT analysis. Ukraine Association Biobank

References

1. European Commission. (2012). *Biobanks for Europe. A challenge for governance.* Retrieved from https://op.europa.eu/en/publication-detail/-/publication/629eae10-53fc-4a52-adc2-210d4fcad8f2
2. Cañada, J. A., Tupasela, A., & Snell, K. (2015). Beyond and within public engagement: A broadened approach to engagement in biobanking. *New Genetics and Society, 34*(4), 355–376. https://doi.org/10.1080/14636778.2015.1105130
3. Bjugn, R., & Casati, B. (2012). Stakeholder analysis: A useful tool for biobank planning. *Biopreservation and Biobanking, 10*(3), 239–244. https://doi.org/10.1089/bio.2011.0047
4. Gaskell, G., & Gottweis, H. (2011). Biobanks need publicity. *Nature, 471*(7337), 159–160. https://doi.org/10.1038/471159a
5. Lesch, W., Schütt, A., & Jahns, R. (2016). Biobanken in der öffentlichen Wahrnehmung: Verständnis, Interesse und Motivation von Probenspendern in Deutschland. *Gesundheitsforschung Kommunizieren, Stakeholder Engagement gestalten*, 113–124. https://doi.org/10.32745/9783954663637-3.1.
6. Mitchell, D., Geissler, J., Parry-Jones, A., Keulen, H., Schmitt, D. C., Vavassori, R., & Matharoo-Ball, B. (2015). Biobanking from the patient perspective. *Research Involvement and Engagement, 1*(4). https://doi.org/10.1186/s40900-015-0001-z
7. Simeon-Dubach, D., & Watson, P. (2014). Biobanking 3.0: Evidence based and customer focused biobanking. *Clinical Biochemistry, 47*(4–5), 300–308. https://doi.org/10.1016/j.clinbiochem.2013.12.01

8. Caixeiro, N. J., Byun, H. L., Descallar, J., Levesque, J. V., de Souza, P., & Soon Lee, C. (2015). Health professionals' opinions on supporting a cancer biobank: Identification of barriers to combat biobanking pitfalls. *European Journal of Human Genetics, 13*, 626–632. https://doi.org/10.1038/ejhg.2015.191
9. Buse, K., Mays, N., & Walt, G. (2012). *Making health policy* (2nd ed.). Open University Press.
10. Arnstein, S. R. (1969). A ladder of citizen participation. *Journal of the American Institute of Planner, 35*(4), 216–224. https://doi.org/10.1080/01944366908977225
11. Friedman, A. L., & Miles, S. (2006). Stakeholders. Theory and practice. *Journal of Risk Research, 11*(6), 841–843. https://doi.org/10.1080/13669870701566607

Biobanking IT Systems, Database Structure and Web Applications

12

Christine Mitchell, Svetlana Gramatiuk, Tamara Sarkisian, Zisis Kozlakidis, and Karine Sargsyan

Abstract

The rising importance of biobanks for biomedical research cannot be overstated. They are expected to deliver a sustainable impact on research by generating a high-quality resource with clinically annotated data corresponding with relevant biological samples. This corresponds with the growing demand for biobanks to deliver high-quality samples and, more importantly, high-quality corresponding data and accurate knowledge of sample locations and life cycle (storage method, temperature and its deviation). With the large use of new large-scale analytical technology platforms (-omics), the required capacity of stored data increased exponentially, resulting in greater challenges for the architecture of biobank information management systems and their interoperability for further downstream research. Through such a kind of system, it is possible to control the individual samples and their location at any time on the way to the laboratory and storage capacities: during storage in tanks, on the extraction bridge, during

C. Mitchell
International Biobanking and Education, Medical University of Graz, Graz, Austria

S. Gramatiuk
Institute of Cellular Biorehabilitation, Kharkiv, Ukraine

T. Sarkisian
Department of Medical Genetics, Yerevan State Medical University, Yerevan, Armenia

Z. Kozlakidis
International Agency for Research on Cancer, World Health Organization, Lyon, France

K. Sargsyan (✉)
International Biobanking and Education, Medical University of Graz, Graz, Austria

Department of Medical Genetics, Yerevan State Medical University, Yerevan, Armenia

Ministry of Health of the Republic of Armenia, Yerevan, Armenia
e-mail: karine.sargsyan@medunigraz.at

81

aliquoting or during transport to the workstation or to analysis platforms. These are particularly important requirements for efficient work and the correct use of analytical data and materials for future medical research, with the potential to improve the quality of ongoing work and integrate new functions.

Keywords

SMART · LIMS (laboratory information management system) · Data · Databases · Interface · Database · Data model · Oracle · Web interface · Data management · User management · IT system

Although biobanks play an important role in establishing links between important basic biomedical research and clinical trials [1], the access to high-quality and well-characterized biological resources remains a serious problem. The adequate availability of samples, especially, e.g. in rare diseases, requires the cooperation of several biobanks, which must possess an interconnection and have a similar quality management [2]. The choice, implementation and the use of an IT management system for biobanks is recognized worldwide not only for their biological resources over their lifetime but also for improving data quality and linking biological samples to relevant clinical data. Biobanks that continue to use other "expired" methods of data collection (e.g. spreadsheets, paper records and handwritten storage tubes) are in very poor condition. With the increasing use of large analysis platforms, the amount of data generated has exponentially increased due to its relatively affordable value. This posed new challenges for the BIMS (biobank information management system) architecture. In addition, improving the quality of biological samples stored in the repository will de facto improve the quality of research results and correspondingly help reduce non-reproducible research data, which, according to Friedman et al. [3], is responsible for more than 60% of all nonreproductive causes.

In addition to the data and sample information management, design as well as input, it is important to cover the activities of the specific biobank and the functions that come with its profile. Also, a functioning IT system (software and hardware) strengthens the partnership with all partners.

The growing importance of repositories for research should not be underestimated. In addition, there is a growing demand for biobanks that provide high-quality samples. The ISBER Best Practices for Biobanks and the International Agency for Research on Cancer (IARC) General Minimum Technical Standards and Protocols [4, 5] emphasize the use of an effective biobank IT management solution that tracks efficient biological resources throughout their life cycle. Paskal et al. and Bendu et al. [6, 7] define the use of the information management system as an important subject for the operation of biobanks and laboratories.

The ever-increasing demand for the research warehouse capacity should neither be over- nor underestimated. This forces biobanks to focus on providing high-quality samples rather than on how the samples should be stored. Paskal et al. and Bendu et al. [6, 7] define an efficient IT system as a key issue for biobanks. They point out that the choice of a particular BIMS is closely related to a number of important properties that will differentiate the analysis results for each repository and service, which is provided by the biobank. There are many BIMS models available from a conceptual point of view. Biobanks provide their own solutions or cloud software, commercial software and open-source software or special laboratory systems with biobank management modules. In developing countries, the most common option is the development of an internal solution, which often requires significant development and maintenance costs [5].

As part of the biobank development project, an evaluation for the decision internal development versus commercial BIMS has to be performed based on selected requirements. Among others, the following criteria must be taken into account:

1. Basic functionality
2. Reports and barcode reading/printing
3. Quality, safety and compliance
4. Sample inventory function
5. Data management (speed and security)

Currently, there is no exhaustive list of BIMS, so each biobank, depending on its profile and requirements, must conduct a market analysis to find a suitable management solution [8]. Most BIMS are derived from laboratory information management systems and perform a number of key functions [9]. The main activities relate to the registration and monitoring of biological resources (biological samples, relevant clinical, demographic or laboratory data) from their receipt to their subsequent distribution. The importance of keeping track of the laboratory's progress, organizing the data from subsequent analyses and reviewing the recorded data cannot be underestimated. All of these testimonials contribute to the long-term value of the recorded sample.

The BIMS is indeed a powerful tool associated with a large number of samples in biobanks. It contributes to the improvement of the quality of biological resources, ensures the development of biobanks (methods, infrastructure, consumable change, etc.) and keeps every step recorded. In addition, it enables complete consistency in the supply chain by tracking every step from sample to delivery, reducing potential errors and increasing the communication capacity [5, 9, 10]. From an IT perspective, each biobank process is an important workflow in itself, so the BIMS supports a number of processes that make it easier for biobank staff to access and manage samples. Biobank databases are custom configurations designed to better support a specific storage process by addressing the requirements of the specific biobank workflow in the database. BIMS can keep track of who processes each sample at each stage of the workflow [10] for all samples stored in the repository, which helps each biobank with their quality assurance and control.

To determine the BIMS requirements, each biobank must follow their own processes to document the exact needs. Biobank scientific organizations have already made significant efforts to standardize data collection and facilitate the exchange of data between biobanks. There are several papers and other scientific literature that focus on the BIMS implementation concepts and requirements [11–16]. The biobanking and biorepository standards [4, 5, 17–22] were also subsequently revised to meet the requirements of modern biobanking and the development of analytical technology. These developments and revised standards were based on biobank business plans, publications and discussions on institutional restrictions or regulations, legal advice and best practice guidelines. In addition, brainstorming techniques were used, and unstructured expert group discussions not only in biobanking but also with laboratory sciences or with pathologists were carried out to classify laboratory-related requirements of the biobank workflows.

The selection process for defining an IT solution can be very extensive and can take more time and effort as initially expected. The identification of affordable IT solutions on the market can be done in consultation with the national network of biobanks, local collaborative platforms, etc., taking into account suppliers known in the field of biobanks or active in laboratory IT. Given the possibility of a common doctrine in a specific country, region or multinational organization, overhead regulations will have a greater impact on negotiations with future service providers and will be facilitated by the use of the same IT system [23]. The first assessment must be performed by analysing the number of free features published by the provider on their website, to have a short list for further work and to be prepared for securing data lifecycle management in research [7, 24]. The next round of selection should focus on the live demonstrations of exposed systems. User requirement cases (use cases) must be identified, briefly described and provided to the developing team or commercial suppliers for the extensive assessment. Previously defined short requirements can be used as a checklist to assess the effectiveness and adherence of processes.

To enhance and support the operation and the service of a biobank, the IT solution should support the registration of all donors of the biobanks. Due to strong ethical and legal requirements in the biobanking field, the patient or donor may terminate the contract at any time without delay. All IT systems and programmes in the biobank must include processes to support the immediate patient/donor withdrawal. Also, the biobanker should be able to see all the stored samples and the available data with the corresponding usage possibilities and restrictions (if any).

The creation of the databank architecture for a biobank should follow the relational design process. Data to be recorded in the database must be identified and described (as far as possible) in the specification book from the outset. Specific donor and biospecimen criteria must be defined, but also several standard donor identification categories (e.g. ID, patient, institute) must be included to facilitate the possible interconnections with the hospital information system and/or analytic platforms. The linking of data and its associations should be implemented for completeness of the data. The primary keys (patient identification data, represented in numbers) represent unique codes that are unique to each record and should be used to match a dataset or biospecimen to a donor and to establish relationships between tables in the databank, i.e. between different pieces of information. The creation of the relationship between different tables of the databank should be performed by the IT team (own team or commercial provider), but the concept must be done by the management of the biobank and the project management group to fully support the biobank routine.

Non-profit research biobanks are often developed by scientists for the needs of their own research or the research of their own organization, with little or no IT project management skills. Thus, this work can be supported by several publications, such as Prokosch et al. [12], who offer a process that demonstrates the complexity for a research biobank to implement a BIMS in conjunction with its own institution's IT department. The main challenge for non-profit biobanks is the implementation of management-oriented processes with little or no knowledge of IT-specific project management.

For a database plan and smooth implementation, the first important task is to define the company's goals by analysing the processes and defining the benefits of implementing such an IT system. These objectives will guide the biorepository through the selection process. BIMS compliance with safety and regulatory requirements is of great importance and a crucial point in the decision (especially for an internal or external system). The traceability of stored biological resources (sample and clinical annotations) has to be ensured.

The IARC, in its technical publication No. 44 [5], defines that "IT systems must ensure complete traceability of samples and data" and that "it can be accessed only by authorized personnel". During workflow analysis, a biobank should demonstrate evidence of scalability, the ability of systems to be modified in the future, business-specific requirements, and upgradeability, as well as the likelihood of future update support and, in the case of new device implementation, connectivity. In addition, the usability (user-friendliness) of the system is the essential prerequisite for its acceptance by biobank employees. It will contribute significantly to the biobank employees' decision to use it. Therefore, intuition and the intended training time to acquire basic knowledge must also be factored in.

In our opinion, support and cost must be weighted as factors with the second highest score. For a small biobank without IT support from the institution, the services that the vendor can provide are crucial for a smooth system implementation. It is important to identify whether the vendor offers live support and annual service plans, and whether licencing costs cover major system upgrades and integration support for workflow automation development.

Furthermore, the system acquisition and/or development costs have a major impact on the sustainable development of the biobank. Providers have different business strategies, and the offers are sometimes not comparable. It is important not only to determine the upfront budget (target price) but also to calculate and include license fees, configuration and/or customization work, implementation and integration support, quality and validation test assessment, etc. Clearly, the quality and complexity of the implementation will have an impact on the final expenses. In addition, the acquisition of an out-of-the-box (specially customized) system, combined with a phased approach that prioritizes the requirements with the greatest return on investment in the first phase, has a positive impact on the sustainability of a small- to medium-sized repository, as described by Nussbeck et al. [13].

For short-term developments and effects, there are tips for optimal use of the BIMS search engine, which has been reported by Kyobe et al. [25]. Nevertheless, criticisms must be made of the development regardless of how sophisticated and complete the programme and its logic are, and all possible options to facilitate an information solution that meets most of the resources needed must be carried out.

The IT landscape of UAB is presented as an example of an in-house implementation.

The laboratory and biobank IT programme—SMART—is representing the most technically advanced laboratory information management system in the Ukraine from the time when it was introduced to the market until now.

The SMART system has been developed together with a UAB specialist to serve also the biobanking needs in the Ukrainian market. It is a browser-based deployment with web interfaces and a flexible and configurable web service. It contains a rich catalogue of attributes, which can be modularly implemented in a laboratory and a biobank serving the required management functionality.

The first prerequisite of SMART was to link it to all storage capacities independent of the typical inconsistency and heterogeneity of storage unit categories and temperatures. At the beginning of the implementation of SMART, in the UAB member biobanks and sites, a survey was conducted with information on collections and facilities. This provided a basis of definitions and detailed specifications including the wide range of the UAB members' sample diversity. The structure of storage containers is a reflection of the association's infrastructure such as the following three types: humidity-controlled room temperature storage, nitrogen reservoirs and freezers. A wide-ranging and fixed chain of links (from the tube to the box to the rack to the stage, etc.) describes the exact location of each aliquot. Precise general as well as rest capacity for the storage unit is documented in SMART for capacity monitoring purposes of each partner biobank. The hierarchic structure in the SMART system is stiff; nevertheless it can grow integrating new ranks, new storage units or transfer of samples within the unit (defragmentation).

The second prerequisite for SMART was the constant tracking of the samples and the documentation of all actions in the system, including the management of the active procedures from specimen retrieval to the in-house analytical measurements and the shipping.

The third prerequisite for SMART was the fast import and export of the corresponding data from medical records and various databases and to be able to input the information for any certain sample collection. SMART does not substitute for the location or collection type-driven databases but rather relies on these databases for flexible administration and control.

As a final point, SMART validated all security controls to fulfil the requirements of data safety and confidentiality in accordance with the ethical standards of the UAB.

In Fig. 12.1, it is shown how SMART manages the requests and the functions to access big data collections through an easy-to-use and user-friendly interface.

Considering the size of the database, the relational data model is a logical decision, which was developed in Release 2 Oracle (10.2.0.1). The database has

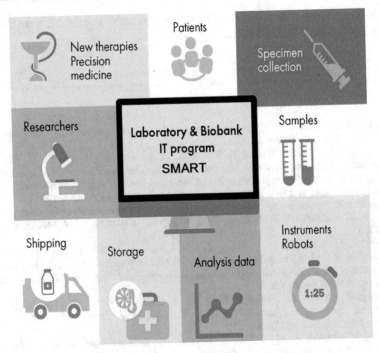

Fig. 12.1 The SMART operates control comprehensive datasets through a sincere user-friendly interface. Ukraine Association Biobank

the ability to manage a large amount of data, which increases the speed of processed requests in the system. The system includes the possibility of a high level of recovery in ensuring ethical and legal security and guarantees for the tools used.

The database model used in the SMART IT system improves the opportunities for greater flexibility provided by the storage facilities. The database model of the UAB includes clinical information, (non)exclusively collected for projects, and the exact identification of the tables used, different samples, aliquots, tube types, containers and storage units.

SMART database management uses graphical and web interfaces developed on 6.0 Oracle Developer in the Java operating module using Java Soft JDK. The independent platform interface can adapt to the user and its role and relation to a member biobank of the UAB depending on the programmable user management.

The system allows to control the single specimen and its place on the path of laboratory and storage capacities everywhere: when storing in tanks, at extraction bench, by aliquotation or transportation to the workplace or to analytic platforms.

The extensible SMART model is the most promising model of IT systems in medicine and biobanks in the Ukraine. This concerns especially its key requirements for effective work and the correct use of analytical data and materials for future medical research with the possibilities of improving the quality of ongoing work and integrating new features. The SMART database is adapted to a variety of collections and different storage structures. This system can assist any biobank or research centre and is usable with all known web interfaces.

References

1. springer.com. (2019). *Advances in experimental medicine and biology*. Retrieved May 25, 2021, from https://www.springer.com/series/5584
2. Mate, S., Kadioglu, D., Majeed, R. W., Stöhr, M. R., Folz, M., Vormstein, P., Storf, H., Brucker, D. P., Keune, D., Zerbe, N., Hummel, M., Senghas, K., Prokosch, H. U., & Lablans, M. (2017). Proof-of-concept integration of heterogeneous biobank IT infrastructures into a hybrid biobanking network. *Studies in Health Technology and Informatics, 243*, 100–104.
3. Freedman, L. P., Cockburn, I. M., & Simcoe, T. S. (2015). The economics of reproducibility in preclinical research. *PLoS Biology, 13*(6), e1002165. https://doi.org/10.1371/journal.pbio.1002165
4. Campbell, L. D., Astrin, J. J., DeSouza, Y., Giri, J., Patel, A. A., Rawley-Payne, M., Rush, A., & Sieffert, N. (2018). The 2018 revision of the ISBER best practices: Summary of changes and the editorial Team's development process. *Biopreservation and Biobanking, 16*(1), 3–6. https://doi.org/10.1089/bio.2018.0001
5. Mendy, M., Caboux, E., Lawlor, R. T., Wright, J., & Wild, C. P. (2017). *Common minimum technical standards and protocols for biobanks dedicated to cancer research*. International Agency for Research on Cancer.
6. Paskal, W., Paskal, A. M., Dębski, T., Gryziak, M., & Jaworowski, J. (2018). Aspects of modern biobank activity—Comprehensive review. *Pathology Oncology Research, 24*(4), 771–785. https://doi.org/10.1007/s12253-018-0418-4
7. Bendou, H., Sizani, L., Reid, T., Swanepoel, C., Ademuyiwa, T., Merino-Martinez, R., Meuller, H., Abayomi, A., & Christoffels, A. (2017). Baobab laboratory information management system: Development of an open-source laboratory information management system for biobanking. *Biopreservation and Biobanking, 15*(2), 116–120. https://doi.org/10.1089/bio.2017.0014
8. Kersting, M., Prokein, J., Bernemann, I., Drobek, D., & Illig, T. (2014). IT-Systems for Biobanking—A brief overview. Im, K., Gui, D., Yong, W. H. (2019). An introduction to hardware, software, and other information technology needs of biomedical biobanks. *Methods in Molecular Biology (Clifton, N.J.), 1897*, 17–29. https://doi.org/10.1007/978-1-4939-8935-5_3
9. Im, K., Gui, D., & Yong, W. H. (2019). An introduction to hardware, software, and other information technology needs of biomedical biobanks. *Methods Molecular Biology (Clifton, N. J.), 1897*, 17–29.
10. Vaught, J. B., & Henderson, M. K. (2011). Biological sample collection, processing, storage and information management. *IARC Scientific Publications, 163*, 23–42.
11. Quinlan, P. R., Mistry, G., Bullbeck, H., Carter, A., & Confederation of Cancer Biobanks (CCB) Working Group 3. (2014). A data standard for sourcing fit-for-purpose biological samples in an integrated virtual network of biobanks. *Biopreservation and Biobanking, 12*(3), 184–191. https://doi.org/10.1089/bio.2013.0089
12. Prokosch, H. U., Beck, A., Ganslandt, T., Hummel, M., Kiehntopf, M., Sax, U., Uckert, F., & Semler, S. (2010). IT infrastructure components for biobanking. *Applied Clinical Informatics, 1*(4), 419–429. https://doi.org/10.4338/ACI-2010-05-RA-0034
13. Nussbeck, S. Y., Skrowny, D., O'Donoghue, S., Schulze, T. G., & Helbing, K. (2014). How to design biospecimen identifiers and integrate relevant functionalities into your biospecimen management system. *Biopreservation and Biobanking, 12*(3), 199–205. https://doi.org/10.1089/bio.2013.0085

14. Eminaga, O., Semjonow, A., Oezguer, E., Herden, J., Akbarov, I., Tok, A., Engelmann, U., & Wille, S. (2014). An electronic specimen collection protocol schema (eSCPS). Document architecture for specimen management and the exchange of specimen collection protocols between biobanking information systems. *Methods of Information in Medicine, 53*(1), 29–38. https://doi.org/10.3414/ME13-01-0035

15. Späth, M. B., & Grimson, J. (2011). Applying the archetype approach to the database of a biobank information management system. *International Journal of Medical Informatics, 80*(3), 205–226. https://doi.org/10.1016/j.ijmedinf.2010.11.002

16. Schwanke, J., Rienhoff, O., Schulze, T. G., & Nussbeck, S. Y. (2013). Suitability of customer relationship management systems for the management of study participants in biomedical research. *Methods of Information in Medicine, 52*(4), 340–350. https://doi.org/10.3414/ME12-02-0012

17. Betsou, F., Lehmann, S., Ashton, G., Barnes, M., Benson, E. E., Coppola, D., DeSouza, Y., Eliason, J., Glazer, B., Guadagni, F., Harding, K., Horsfall, D. J., Kleeberger, C., Nanni, U., Prasad, A., Shea, K., Skubitz, A., Somiari, S., Gunter, E., . . . International Society for Biological and Environmental Repositories (ISBER) Working Group on Biospecimen Science (2010). Standard preanalytical coding for biospecimens: Defining the sample PRE analytical code. *Cancer Epidemiology, Biomarkers and Prevention: A Publication of the American Association for Cancer Research, Cosponsored by the American Society of Preventive Oncology, 19*(4), 1004–1011. doi: https://doi.org/10.1158/1055-9965.EPI-09-1268.

18. Lehmann, S., Guadagni, F., Moore, H., Ashton, G., Barnes, M., Benson, E., Clements, J., Koppandi, I., Coppola, D., Demiroglu, S. Y., DeSouza, Y., De Wilde, A., Duker, J., Eliason, J., Glazer, B., Harding, K., Jeon, J. P., Kessler, J., Kokkat, T., Nanni, U., . . . International Society for Biological and Environmental Repositories (ISBER) Working Group on Biospecimen Science (2012). Standard preanalytical coding for biospecimens: Review and implementation of the sample PREanalytical code (SPREC). *Biopreservation and Biobanking, 10*(4), 366–374. https://doi.org/10.1089/bio.2012.0012.

19. Norlin, L., Fransson, M. N., Eriksson, M., Merino-Martinez, R., Anderberg, M., Kurtovic, S., & Litton, J. E. (2012). A minimum data set for sharing biobank samples, information, and data: MIABIS. *Biopreservation and Biobanking, 10*(4), 343–348. https://doi.org/10.1089/bio.2012.0003

20. BIMS Guidelines. (2021). *Swiss biobanking platform.* Retrieved May 25, 2021, from https://swissbiobanking.ch/specification-for-a-bims/

21. Data Collection. (2018). *EBMT.* https://www.ebmt.org/registry/data-collection

22. BBRB. (2018). *Vocabularies for biobanking.* Retrieved May 25, 2021, from https://biospecimens.cancer.gov/resources/vocabularies.asp

23. 20/20 Vision—HUG—Making the HUG Efficient, Welcoming and Effective. (2019). *Hôpitaux Universitaires de Genève.* Retrieved May 25, 2021, from https://www.hug.ch/en/vision-2020

24. LIMSWiki. (2020). *LIMSWiki.* Retrieved May 25, 2021, from https://www.limswiki.org/index.php/Main_Page

25. Kyobe, S., Musinguzi, H., Lwanga, N., Kezimbira, D., Kigozi, E., Katabazi, F. A., Wayengera, M., Joloba, M. L., Abayomi, E. A., Swanepoel, C., Abimiku, A., Croxton, T., Ozumba, P., Thankgod, A., Christoffels, A., van Zyl, L., Mayne, E. S., Kader, M., & Swartz, G. (2017). Selecting a laboratory information management system for biorepositories in low- and middle-income countries: The H3Africa experience and lessons learned. *Biopreservation and Biobanking, 15*(2), 111–115. https://doi.org/10.1089/bio.2017.0006

Management Model and Sustainability Plan of Biobanks

13

Gabriele Hartl, Berthold Huppertz, and Karine Sargsyan

Abstract

It is important to realize that understanding important issues such as the financial and economic situation of UAB member biobanks, which are directly related to biobanking and biobank maintenance, is essential for the successful use of private and public support. The process of developing a biobank cost model in the Ukraine has been studied over many years, taking into account the issue of funding mechanisms to ensure biobank stability. This chapter examines the objectives of the Ukrainian biobank sustainability plan. The development of an in-depth and detailed research and growth strategy for the UAB was the cornerstone of the sustainability plan.

Keywords

Sustainability · Management tool · Cost recovery · Funding · Sustainability plan

A lesson from the Ukraine

G. Hartl (✉)
International Biobanking and Education, Medical University of Graz, Graz, Austria
e-mail: gabriele.hartl@medunigraz.at

B. Huppertz
Division of Cell Biology, Histology and Embryology, Gottfried Schatz Research Center, Medical University of Graz, Graz, Austria

K. Sargsyan
International Biobanking and Education, Medical University of Graz, Graz, Austria

Department of Medical Genetics, Yerevan State Medical University, Yerevan, Armenia

Ministry of Health of the Republic of Armenia, Yerevan, Armenia

Understanding important issues such as the financial and economic situation of the UAB member biobanks is crucial for the successful use of private and public support.

The process of developing a cost model of biobanking in the Ukraine was examined, considering the question of financing mechanisms (e.g. cost recovery models) to ensure the stability of biobanks over many years.

The purposes of the Ukrainian Biobank sustainability plan are:

- To develop and accomplish an extensive sustainability plan that permits to manage the immediate economic needs for the short-term strategies and the long-term financial and functional sustainability with the support of the focused elaboration and uprated processes. The strategy of the business is detailed.
- Achieve the required financial capacity with the help of the stabilization of the current economic limits and expanding opportunities for a better industry cooperation including not only the biospecimens but also analytical services.

The development of an extensive and detailed research and growth strategy for the UAB was the keystone for the sustainability straining. After completing the development, the strategic plan, which consists of two phases, was presented and considered in full and in detail to the main stakeholders of the UAB, such as the medical institutes and medical universities and the faculties of medicine. This process helped to review and clarify the executive needs.

Included in the content research were the development and the implementation. UAB tried to study more on the economic circumstances of biobanking, especially in low-income countries, and to gain real costs and the possible pricing information, to gain expertise in small market research (Internet research, etc.). The focus was on recovery and funding of the costs, biobank demographics, pricing and specimens as well as products and services.

Some freely accessible industry reports from investment firms about biobanking were an important resource of information for the clear understanding of the market. The research, which is announced by such companies, pointed out the global requirement on tissue, and the accompanying information and study services. In 2015, it was valued at around 850 million euros, and the market growth of biobanks was around 20% per year. It was also mentioned that the biobanking market representation is extremely competitive but is a segmented market.

The team of the UAB carried out a rigorous investigation of operating costs related to the direct market of the Ukraine, both with and without personnel costs. Personnel expenditures were made up of temporary and variable costs. These costs have been discovered and documented in all the aspects of biobanking: starting from the realization of the patient consent, the application and administration of IRB issues, the biospecimen collection as well as the accomplishment and storage, the administration of data, the maintenance, the equipment and possible growth investments.

After identifying the costs, the team at the UAB improved benefit forecasts for the following 5 years, containing all possibilities, which rise from industry and academic research agreements, donor gifts, grants and departments. All these additives increase in value of the UAB and were adapted well disposed to the significant biospecimen disposition instruction for 5 years.

The UAB member organizations (who responded to the survey) indicated that the highest percentage of responses received from biobanks came from government medical (academic) centres (69.9%), municipal clinics (12.9%) and self-regulating scientific laboratories (8.6%). The lowest amount of answers came from industrial-commercial institutions (4.3%) as well as from specific government-supported projects (4.3%). While respondents have been requested information on biorepositories' administrative-structural models, the mainstream showed information on biorepositories, which were dedicated to research actions and performance (66.3%). 17.2% were included in network structures (numerous repositories), 12.9% of the biobanks worked autonomously and 8.6% remained associated biobanks (network).

The funding of the UAB members was defined with probable spreading of the financial sources per year (see Fig. 13.1). The most popular sources of finance were identified to be payments of sample and data usage fees, which have recuperated from tasks of biorepositories. Second popular was the governmental or so-called economic support of non-profit organizations as well as internal funds from carrier organizations. Subsidies in the form of private grants, assistance and generous donations have been described only very rarely.

An original specification to the UAB structure of the human biology collection was created as a so-called fee-for-service model, which ties in with the results of the fully established and flexible personnel and consumables fee and all other costs. A specific market analysis for the Ukrainian market of biobanking showed that these costs are slightly higher than the costs of the competitors. According to the judge-ment of the UAB, it was important to keep a competing degree for an insistence of space on the market, and for more specimen usage from the UAB. A revision of cost calculation with a focus on process optimization and cost reduction was based on the dynamics of the extreme standard construction set-off to be constructed. It gave the conclusive level of self-costs of less than the standard supported market costs (approximately 27% lower), which provided the association with an opportunity to survive on the market.

The UAB drew up a sustainability plan for a building, which initially focused on two questions:

1. Are there any resource opportunities to be gained, and if so, can they be the solution of the funding problem and where should it be addressed to?
2. Is there a possibility of leveraging university-wide or host organization wide resources for biobank development? Can this strategy be an opportunity to be successful? Should the UAB fill the place with a core resource established within a health school?

An extensive UAB project team has supported the implementation of the newly developed sustainability plan, being further part of the UAB to develop inside

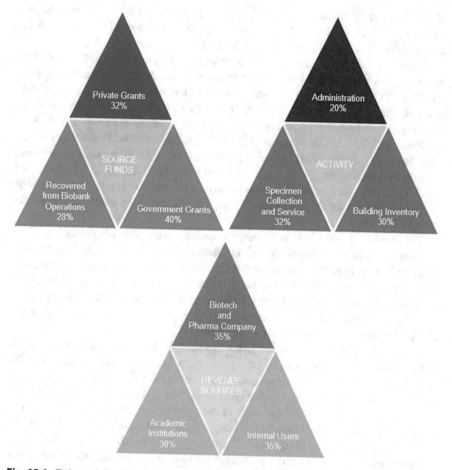

Fig. 13.1 Estimated distribution of funding sources annually. Ukraine Association Biobank

knowledge and the basis for stakeholders. The list of involved people in this innovative development to join the effort consists of existing personnel from various member organizations of the UAB and an MBA (legal) from a regional institute.

Sustainability Plan and Project Management of a Biobanking Network

14

Svetlana Gramatiuk and Gabriele Hartl

Abstract

A planned management approach was used to provide the basic framework for developing, implementing and evaluating sustainability efforts. This approach was particularly useful in gaining insight into the large number of host organisations and stakeholder organisations within the UAB membership. It is important that the work plan is disclosed to all UAB members, setting goals, showing progress, suggesting necessary steps for better alignment and optimising routine work as much as possible. This chapter explains the process of developing a sustainability plan for UAB. Fulfilment of this strategic development plan with the bridging funding shown will lead to short-term financial stabilisation of UAB. The long-term success and continuation of the plan will be assessed based on the metrics achieved and how exactly it works.

Keywords

Strategic planning · Sustainability plan · Cooperation partners · Collaborative investigations · Material transfer agreement (MTA)

A planned management approach was used to create the essential framework to this sustainability effort's development, implementation and evaluation. This approach was especially useful for providing insights to the high quantity of host organisations and stakeholder organisations within the UAB members. Most of the team members

S. Gramatiuk
Institute of Cellular Biorehabilitation, Kharkiv, Ukraine

G. Hartl (✉)
International Biobanking and Education, Medical University of Graz, Graz, Austria
e-mail: gabriele.hartl@medunigraz.at

were not controlled by the host organisation. It is essential that the work plan is disclosed to all members of the UAB, which reveals objectives, makes progress, recommends necessary steps for better alignment and improves routine work as much as possible. It was also important to inform all stakeholders, not just those who were worried. The fact that a project manager was appointed who had an extensive instructional background in the field of biobanking was extremely valuable in this situation. Both extensive experience and institutional knowledge have enabled greater capacity, strength and speed in a plan development. It was also supported by a funded request of evidence designed to secure employees. The UAB realised that settling the parameters of cost reimbursement at the UAB was necessary, but in order to work out, there was the important need for cooperation partners and research projects. The UAB's own fundraised research was also used to achieve a stable income from business activities even to be non-profit, to be able to maintain biobanking in general and develop facilities.

After strategic planning of possible sustainability was composed, improved and scrutinised through the team of the UAB, it was implemented in all member organisations for garnering the internal funding. Specific funding for a period of 1 year (apart from in-kind contributions from the host organisations) was provided by some interested research organisations with the support of the benefiting endowment. Several operational enhancements have been made, also including the system implementation of ISO 9001:2015 and setting up a second-rate constitution structure to intensify security and safety within the UAB members.

The sustainability plan defined and systematically sharpened strategies, which were implemented in two stages: at the beginning, monetary stability (Phase I), followed by centralised manoeuvres in the financial, employment and trade process performance of the UAB (Phase II). The timelines were improved. Figures 14.1 and 14.2 show some specific tactics that were necessary for the consideration of strategy in order to become stable as well as advance a long-term economic capacity of a research biorepository of networks in the Ukraine such as the UAB. One of the main issues was to engage members in construction of industrially supported research with the pharma and biotech industry, as well as implement a standard material transfer agreement (MTA) to be able to work with for-profit and not-for-profit entities, and keep out of the opinion connected with the biospecimens "selling". A close collaboration with the health school, medical institute, medical university and other academic research collaborations positively influenced the situation and was essential for the development.

An evaluation was made for both in-house and accidental investigation partnerships, and the UAB recently released the primary list for an investigation plan including brain tissue. For some additional revenues, other plans are at the interlace stage of agreement. A relationship with a biomarker investigation agent showed results in providing some unique and typical interest in opportunities for collaborative investigations, goals in cancer studies, to the inside and unintentional researchers, patients and the regional participation and charging with revised MTA.

The UAB working group is working and setting out a plan of presenting the Ukrainian biobanks' specific and unequalled characteristics, aware of cooperative

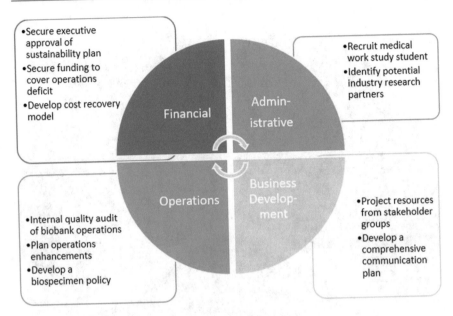

Fig. 14.1 Tactics stabilisation. Ukraine Association Biobank

study chances, as well as goals in cancer investigations to the in-house and peripheral researchers and the regional participation. These home-grown adverts recommend the usefulness of biospecimens among academic institutions and the industry. The team also targeted at some well-known official events, such as popular meetings and conferences, to intensify requests for specimens.

The UAB's sustainability plan was primarily used for garnering support of private grants and research funds, and secondly outlining both short- and long-term operating- and economic modifications. The forceful strategy installed reliability through understanding and describing a need of research and outlining the UAB's value being a biobanking research strength in the Ukraine.

With the help of the method "Milestone—focused, project management" it was proven that documented advances also convince critical stakeholders. Perspective standardisation and cooperation with European biobanks have given an opportunity to biobankers to learn and to develop structures. Also, the opportunity to researchers was given to admit specimens from a network of biorepository collections.

In the meanwhile, the accomplishment of a strategic development plan with stopgap funding showed results in a short-term financial stabilisation of the UAB. Its long-term achievement and succession will be estimated based on the metrics accomplished and the accurate way of working.

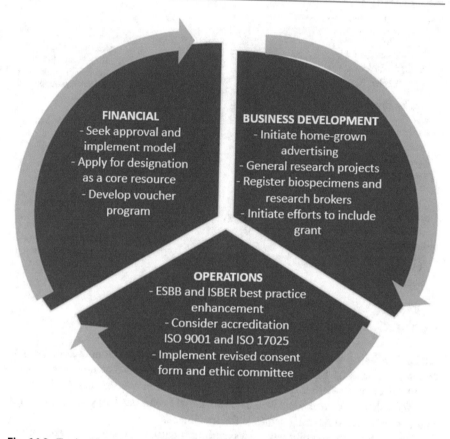

Fig. 14.2 Tactics Financial Solvency. © Ukraine Association Biobank

Establishing a Biobank Using Standardisation

15

Mykola Alekseenko, Christine Mitchell, and Svetlana Gramatiuk

Abstract

Interethnic variations in Ukraine may influence the number of chromosomal aberrations detected and the treatment and pharmacokinetics. Studies of different institutions determine the identification of singular oncological diseases typical for Ukraine. The UAB Central Biobank has medical specialists in the institutions, especially in the Central Data Management Department, the Clinical Trial Services Department and the Research Department, which provides highly operational and continuous research support. A Biobank Steering Committee and Medical Advocate have been appointed to support the collection and practical consumption of biospecimens in a multicentre design.

The UAB has applied the ESBB criterion in the SOPs for tissue storage, tanking and processing from the beginning. In general, the SOPs implemented in the network support these standards with minor modifications. Policy coherence and SOP as progress, speed and protocols brought about concerted collaborations of data protection, database management, data sharing, tissue collection and tanking, ethical considerations, sample accessibility and workshops and training.

Keywords

Establishment of a network · Ukraine Association of Biobank · Management Committee · SOP · ESBB · Standards · Commission Ethical and Bio-Ethical (CEB)

An Example from the Ukrainian Biobank Network

M. Alekseenko · S. Gramatiuk (✉)
Institute of Cellular Biorehabilition, Kharkiv, Ukraine

C. Mitchell
International Biobanking and Education, Medical University of Graz, Graz, Austria

© The Author(s), under exclusive license to Springer Nature Switzerland AG 2022
K. Sargsyan et al. (eds.), *Biobanks in Low- and Middle-Income Countries: Relevance, Setup and Management*, https://doi.org/10.1007/978-3-030-87637-1_15

Interethnic variations in the Ukraine may influence the number of chromosomal aberrations detected and may influence medication and pharmacokinetics. Studies of the Institute of General and Emergency Surgery NAMS of Ukraine and Institute of Medical Radiology of S. P. Grigoriev National Academy of Medical Sciences of Ukraine and the Department of pathophysiology of Kharkiv Medical Academy of Post-graduate Education insinuate that undeniable genetic diseases personate a material physical interest for the Ukraine. These embrace genetic oncology, rheumatologic, metabolic, endocrine, neurological, haematological and ophthalmological circumstances, as well as inherited malformations. The hospitals of the Medical Institute also designated the identification of singular oncological diseases, which are typical for the Ukraine.

It is therefore expected that the sectional biobanks will be authoritative for meditation, anticipation and curing oncological diseases. It is also confirmed that they contribute particularized molecular and pharmacological data for expert research.

A network overarching the 23 medical centre managers becomes the project of the Ukraine Association of Biobank. The Central Biobank (ASK-Health, owning the central storage capacities) of the UAB has medical specialists in institutions especially in the central data administration department, the department of clinical trial services (I–IV phases) and the department of research to offer very operative and continuous research maintenance. The Biobank Management Committee and the medical lawyer were appointed to encourage the collection and the hands-on consumption of biospecimens in a multi-centric construction.

The UAB members are increasing the sample collection as well as the biobank service maintenance. Also, a "platform" for the centralization of efforts in the UAB was installed. This generates an arrangement, which encourages a comprehensive, wide-ranged and mutual research with associated medical institutions and hospitals (see Fig. 15.1).

The network of the UAB is important for the region-specific human oncological diseases. The concept suggestion of biobanks for this specific intention must be implemented since biorepositories store collections of human biospecimens. However, in this case the main value is the associated data. Due to economic difficulties in the Ukraine and the lack of state funding of the UAB, a decision was made to create a single central biobank storage facility for all participants and members.

The beneficial progression of any biobank is reciprocal to the general support of the population and the willingness to voluntarily donate. To note that, while the limited data on the Ukraine's public opinion regarding biobanking is positive, little is known about potential concerns of the general public, and data derived from other areas may not be applicable to the Ukraine. They indicate that aside from the educated part, most of the population is unfamiliar with concepts related to genetic research (e.g., DNA, genes). This factor may limit the willingness without initiatives designed to educate the public and encourage a participation.

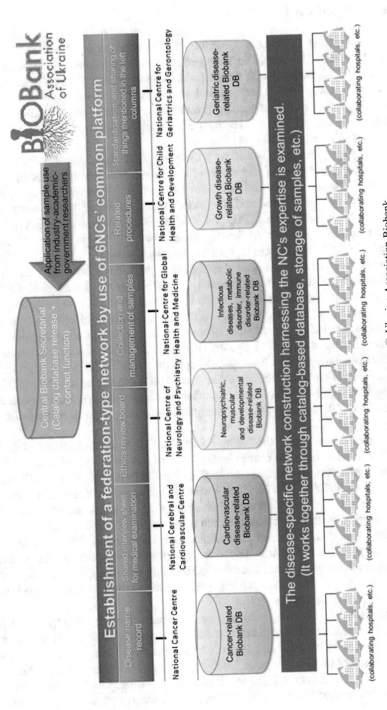

Fig. 15.1 Enterprise of health care improvement for next generation treatment. © Ukraine Association Biobank

Example from UAB

From the beginning, the UAB exercised the ESBB criterion in SOPs to store, tank and handle tissue. The SOPs that were generally exercised across the network are supporting these standards with insignificant modifications. Policy coherence and SOP is a progress, quickness and protocol accomplished harmonious collaborations of workshops and education in data preservation, databank administration, data division, tissue compilation and tankage, ethical considerations, specimen accessibility and quality counteract.

At the outset, it was essential to project an arrangement for the disclosure, revision and approbation of the strategy. It was essential that the policies were conformed with the national and European rules and the directives. It was the first range when the UAB presented the standardization list of the Patients and Volunteers Information Letter (PIL and VIL), the Consent Form Biobank (CFB) and the SAP in the Ukraine. At the beginning, it was difficult to discern the right order of submitting the documents for review in the UAB.

The Irish footpath was recognized as the most qualified one, and it could also be recommended that a similar arrangement of instructions can be adopted for the development and compliance in the Ukraine. First, an agreement with every hospital in the network was needed. Secondly, obtaining a formal legal opinion from an independent medical lawyer after the documents have been reviewed by the Risk and Legal Department at UAB. Thirdly, the Ukraine Association of Biobanks assesses all documentation. Fourthly, the applications to the research and ethics committees of the individual clinics within the biobank network were approved. The fifth step comprised the review of the entire documentation by the data protection commissioner, whereby the commissioner explicitly focused on the proposed method of data sharing (see Fig. 15.2).

During the 5-year period, the National Scientific Centre for Medical and Biotechnical Research has been the central institution for Commission on bioethics at the Cabinet of Ministers of Ukraine (Decision of the Cabinet of Ministers of Ukraine from 13.12.01 № 1677). The latter has encouraged a formation of lawful conditions for linking the Ukraine to the global abridge in bioethics ramification and interconnecting quickness of ethical committees of all kinds in the Ukraine.

In the Ukraine, most Commission Ethical and Bio-ethical (CEB) are under the executive inspection of the Administrative Council of each physical centre. In fact, the course realities of any mind that are persistent and solid direction setters or study exponents and researchers are outlined by elementary groups of study institutions and by the healthcare system.

If one considers, regarding performance, that there are no precise regulations in the Ukraine, which could restrict the establishment and direction of biobanks, one assumes an elementary study plan. The ethical regression of adults is on the scale and is more often left to the concerns of CEBs and their associated institutions. Consent configuration and patient enlightenment pamphlets, which have been projected to beseech some authorizations for biobanking, are particularly modern in the Ukraine.

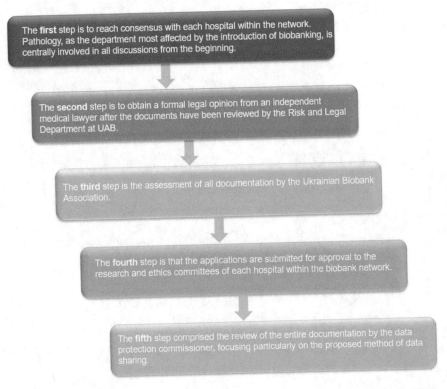

Fig. 15.2 Step-by-step rules on which documents are submitted for consideration in UAB. © Ukraine Association Biobank

For a long time, the patients implored to present blood and/or tissue for a specific study combination, and advertisement appertain to a discriminating plan was prepared.

Science and Innovation of Biobanks

16

Karine Sargsyan, Mykola Alekseenko, Christine Mitchell, and Gabriele Hartl

Abstract

A new era of medical research conveyed numerous innovative discoveries, novel understandings and evidences in the last 20 years and is referred to as the era of personalized medicine. It includes firsthand knowledge and methods but also new patterns of present-day medicine, covering the whole range between prevention and healing. The science of biobanking covers a wide range of scientific fields including research, education, finance, engineering, ethics, quality management, publicity activities, scientific services and others. Many actions have arisen to assist the progress of biobanks. In 2005, the *Office of Biobank and Biological Specimen Research (OBBR)* was founded on the basis of the *US NCI.* In addition, many activities were planned in Europe to encourage the progress of biobanks, some of them under the EU's *Seventh Framework Program (FP7) 2007–2013* and also under the Horizon Framework Program 2020 (2014–2020). EU-funded developments are groundbreaking the progress of genetic methods and carry out projects in large populations on the hereditary predisposition to serious illnesses.

K. Sargsyan
International Biobanking and Education, Medical University of Graz, Graz, Austria

Department of Medical Genetics, Yerevan State Medical University, Yerevan, Armenia

Ministry of Health of the Republic of Armenia, Yerevan, Armenia

M. Alekseenko
Institute of Cellular Biorehabilition, Kharkiv, Ukraine

C. Mitchell · G. Hartl (✉)
International Biobanking and Education, Medical University of Graz, Graz, Austria
e-mail: gabriele.hartl@medunigraz.at

© The Author(s), under exclusive license to Springer Nature Switzerland AG 2022
K. Sargsyan et al. (eds.), *Biobanks in Low- and Middle-Income Countries: Relevance, Setup and Management*, https://doi.org/10.1007/978-3-030-87637-1_16

105

Keywords

Personalized medicine · Precision medicine · Horizon framework program 2020 · Engage · Hypergenes · Gen2Phen

A new era of medical research conveyed numerous innovative discoveries, novel understandings and evidences in the last 20 years and is referred to as the era of personalized medicine. It includes firsthand knowledge and methods but also new patterns of present-day medicine, covering the whole range between prevention and healing. It is also clear that an arrangement of approaches from several disciplines is required to better comprehend health and disease. A new era of biomedical research also brings with it other types of questions that demand answers.

Precision medicine (old term: personalized medicine) is a novel method of diagnosing and treating a patient based on a whole range of elements, such as so-called "omics" technologies (metabolomics, proteomics, epigenomics, etc.), systemic approaches to modern medicine, bioinformatics and, last but not least, biobanks and biorepositories with the associated data structures. The application of precision medicine requires a confluence of several factors [1].

In 2009, biobanks were on the list of "10 Ideas Changing the World Right Now" published in Time Magazine [2]. In this article, the biobank is presented as "a safe stock" of human biological materials, as well as cells, DNA, blood and derivatives—to be used in research activities aimed at innovative diagnostics and treatment for a better life of the Earth's population [2].

Overall, biobanks are particularly flexible because they can assist a diversity of biomedical investigations, such as cross-section projects of genotype-phenotype associations/relationships, case-control investigations that use a biorepository for diseased and/or control material as well as cohort studies that use first time investigation to compare to follow-up with an established link to genetic variation of health outcomes [3].

Biological samples are collected and stored in connection with both epidemiological and clinical research over many decades. The first biobank-like infrastructure facilities have occurred in multiple arrangements for almost 200 years, from tiny collections in the early stages to modern fully automated methods for processing in biorepositories with millions of samples [4]. Thus, it is well known that biobanks as a concept are not new.

The science of biobanking covers a wide range of scientific fields including research, education, finance, engineering, ethics, quality management, publicity activities, scientific services and others. Many actions have arisen to encourage the development of biorepositories and biobanks. In 2005, the *Office of Biobank and Biological Specimen Research (OBBR)* was founded on the basis of the *US NCI*. In addition, many activities were planned in Europe to assist the expansion of biobanks, some of them in frames of the *EU's Seventh Framework Program (FP7) 2007–2013* and others under the *Horizon Framework Program 2020 (2014–2020)*. EU-funded activities are initiating the progress of population-based genetic methods and carry out studies in large populations on the genetic susceptibility and possible disposition to serious illnesses. They also support the development of harmonization procedures and the collection, storage and management of human biological material and genomic information through Europe. Given the importance of population-based methods for studying genetic vulnerability to disorder, the *European Commission's Framework Programs for Research and Technological Development (RTD)* allocated over 60 million euros to joint scientific actions in this area between 2002 and 2008. The most relevant projects will be mentioned here to provide an overview of developments that is as holistic as possible.

The *P3G (Public Population Project)* is an international consortium of representatives from 40 countries. Its purpose is to guide, stimulate and coordinate international efforts and knowledge to enhance the usage of research, biobanks, scientific databases and other comparable health and social investigative infrastructures (http://www.p3g.org).

The *SPIDIA Project (Standardization and Improvement of Generic Preanalytical Instruments and Procedures for* In Vitro *Diagnostics,* http://www.spidia.eu*)* was introduced in 2009 by joint efforts of 16 academic institutions, international organizations and industry actors working in the field of biology. The project contributes to the standardization and improvement of in vitro diagnostic preanalytical procedures [5].

The *ENGAGE consortium* (http://www.euengage.org/) brings together 24 leading scientific institutions and two biotech and pharmaceutical industry representatives in Europe, Canada and Australia. ENGAGE strives to transform the vast amounts of information from large-scale genetic and genomic epidemiological studies of European (and other) populations into relevant information for potential clinical purposes. ENGAGE's model is to empower European scientists to detect a considerable number of novel vulnerability genes that affect metabolic, behavioural and cardiovascular properties and to investigate the relations between genes and lifestyle biomarker factors.

The ENGAGE consortium (http://www.euengage.org) will combine and analyse one of the largest datasets of human genetic data (over 80,000 scans of associations of the entire genome, as well as DNA and blood derivate samples from more than 600,000 persons). One of the goals is to show that ENGAGE results can be expended as novel analytic markers of most common sicknesses that will support us to better recognize risk factors and disease progression and wherefore individuals respond differently to treatment.

The *HYPERGENES project* (http://www.hypergenes.eu) aims to define a complex genetic as well as epidemiological exemplary of multifactor traits, such as hypertension (EH) and phenotypes affecting intermediate target organ damages (TOD) dependent/associated with hypertension, as well as other endophenotypes, as a pharmacogenomic model of drugs commonly used in EH. The recognition of the hereditary factors in common spread illnesses is exceptionally difficult, as over 90% of them are multifactorial and the genetic component is probably described by the relations of several genes convoluted in the pathway, each with a subtle predisposition to disease. HYPERGENS have used a *genome-wide association (GWA)* approach to identify common variants that contribute to a genetic element of common diseases.

The GEN2PHEN project (http://www.gen2phen.org/) *aims* to integrate databases on the genetic variation of human organisms and models to enable an increasingly holistic view of genotype-to-phenotype (G2P) data as well as connect this system with other biomedical knowledge sources through the functions of a genomic browser. At the end of the project, it will create the technological elements necessary for the development of nowadays various G2P databases in a transparent G2P biomedical information setting. It will be a hierarchy centred in Europe, but within the global network of bioinformatic databases, tools and standards connected to GRID, all of which are connected to the browser of the Ensembl genome.

To the best of our knowledge, a total of 34 projects have received support from the *European Commission under FP7 and this process continued in Horizon 2020.*

Not just the European Union framework programs such as *FP5, FP6, FP7 and Horizon 2020* support biobank building and developing activities in Europe and worldwide but also several transnational initiatives also funding the progress of biobanking sciences. The *Innovative Medicines Initiative (IMI,* http://www.imi. europa.eu/), which is still the largest public–private resource in the EU, aims to facilitate the development of better and safer medical products (pharmaceutical measures, diagnostical measures, new schemata for diagnostic and treatment, etc.) for the affected ones. IMI encourages cooperative science and networks of experts from industry and academia to promote pharmaceutical innovation in Europe. IMI is a joint activity of the EU and the Association of the Pharmaceutical Industry—the *European Federation of Pharmaceutical Manufacturers and Associations (EFPIA)* [6].

Despite the progress made, biobanks still face many challenges in the overall biobanking process that need to be addressed [7]. Biobanking lacks harmonization, standards, harmonized terminology, mutual data features and best practices in data collection and sample handling. An accreditation or certification and rating scheme is or should be introduced to acknowledge biobanks with high-quality evidence and to encourage and recognize researchers who create and sustain biobanks [8].

EU policies on open access and open science complicating the method implementation at scientific organisations, which are consistent with apparently conflicting and opposing innovation goals. From one side, there are publications that confirm the claim that intellectual property rights are a substantial obstacle to the progress of science and innovative research, as they impede the open and fast give and take of newly generated knowledge. Parallelly, academic institutions and universities are encouraging scientists to guard the commercialisation prospective of their research results through patents and/or adjacent collaborations with commercial partners to facilitate the transformation of academic findings into products. While the open conversation and knowledge transfer appears to be incompatible with intellectual property fortification, certainly, patent rights and open science are not essentially incompatible. Equally, these two concepts have an objective to increase the impact of research and the benefits of systematic generated knowledge over total disclosure and to encourage further innovation. After all, the patent system is based on the interchange of total disclosure, which allows a time slot of exclusivity. Actually, the legal disclosure prerequisite could make patents even more open than scientific publications, in which researchers can store important evidence for private reasons, such as protecting the prospect of additional science funding [9, 10]. The way patent/human rights may negatively influence the (open) access to generated knowledge, but donors and the patent system are not incompatible with the values and procedures of open access, open science and/or open innovation. As, the current development of donating industrial patents to research institutions allows scientists to influence and expand obtained (protected) scientific data to accelerate research in other areas and foster cross-industry innovation [11]. Open marketing and open access/science as such can be comprehended as corresponding approaches in a coherent innovation framework, aiming to achieve ideal public and economic significance from publicly financed investigations. Different methodologies of innovation policy can be incorporated into a comprehensible framework, which encompasses equally open scientific collaboration and commercial use of scientific data.

The core of the described difficulty is the negative and destructive connotation of the word "commercialization" [12]. The concept of making wealth and profit, in particular the use of missile defence as the main resource, is generally considered to be conflicting with public policy [12, 13]. Nevertheless, in the framework of health sciences and medical research development, commercialization is a very complex concept that includes the concept of access to basic research for the benefit of society. A pragmatic approach and a belief that translation and commercialization are the only processes through which new ideas can be introduced to improve people's health can alleviate some fears and criticisms. Therefore, the focus should be more on maximizing the use of biobanks for the benefit of the patient than on financial goals [14]. The promotion of biomedical research requires the involvement of a wide range of stakeholders, comprising commercial organizations in the field, as none of the singular participants in the innovation process has the abilities and resources for independent research, innovation development, transformation and commercialization [5]. In addition, society's desire to share biobank results and promote health suggests that the distribution and/or payment of benefits can help society to embrace the concept of commercialization [15, 16]. A broader approach to improving human health can go beyond its components, and stakeholders can work together to transform publicly funded research into innovation that also benefits society and promotes advanced academic health research and socio-economic development.

Existing research documents remain fragmented and are far from the overall picture of open science and the policy implications for the concessions available to stakeholders to advance the *Responsible Research and Innovation (RRI)* program. There is literature that calls for a reassessment of the conflict of commercialization at the expense of private and public interests and instead focuses on the production of information [17–19]. This approach requires recognition of the ethical and individual nature of commercial trade in research based on biobank samples and statistics, but with a socialist view of the economic aspects of biobanks, which value knowledge from biology to biological agents. Conflict checking, however, is a theoretical exercise that does not recognize direct interests and does not take into account public and donor interests. The call for the public and sponsors to "think differently" concerning open science and commercial usage of gained knowledge did not calm down or transform the general sensitivity of profit motives and business programs at the overhead of the public interest. The up-to-date open information and publications on innovation and commercialization are strongly business-centric and contributing to the common sensitivity that *Intellectual Property Rights (IPR)* enforcement is largely detrimental [20, 21]. The public must recognize the existence of a conflict in order to overcome the intellectual conflict amongst private and public interests and/or open and closed innovation. Instead, it is practical to mobilize the community so that they know that without their involvement, a progress in biotechnology, health and medical sciences and the invention of life-saving treatments will not be generally possible or, if they decide to gamble, may stop in the innovation process.

Thus, including the *RRI (responsible research and innovation)* principles, the overall innovation framework must emphasize on community participation to guarantee that everyone involved in the innovation process understands that transformation is a complex multi-stakeholder process in which each stakeholder plays a key role in its success. It is not sufficient for all innovation process participants to solely acknowledge their role in the innovation process. The overall framework for innovation should have a mechanism that (a) defines what motivates each stakeholder and (b) allows stakeholders to obtain the particular benefits they assume in return for their input to the invention process. To stimulate contribution, each stakeholder must be rewarded or "considered" for doing so. The form of compensation can vary, since each stakeholder has a different schedule, and profits can come at diverse times throughout the innovation process or be delayed till the innovation appears on the market. Stakeholders understand that the transformation is a fragment of the innovation value chain and understand that they are integral to a broader framework that requires their own input to maximize the opportunity of potentially vital innovation.

A value chain can only function if all participants contribute to a mutual objective. To guarantee contribution, each participant must be able to receive the desired reward. Biobank donors want to take advantage of randomized research. The public expects benefit from the tax money that the government invests in publicly financed basic research. To reap these profits, commercial organisations must transform and standardized early scientific findings gained from biomaterials and/or data from biobanks. Though, the industry will just contribute if there is an economic incentive to do so. For industry to translate, scientists must have the resources to conduct their scientific investigations, rather in an setting encouraging to the creation and dissemination of knowledge. Scientists want free and open access to data to expand scientific knowledge and disseminate results. "Academic entrepreneurs" motivated by their economic interests and acting as intermediaries between science and industry have experience in using external funding for market-oriented research [22]. Universities and public research institutions should provide funding for research-led research. Technology transfer promoted by universities and public and private research and development collaborations bridges the gap between science and practice by pooling competencies to find applications that meet society's needs. The resulting knowledge and innovation in research will contribute to the understanding of science and improve access to funding [23]. The state shall provide public funding for basic research as long as these funds are used to support research that can promote economic growth and does not counteract social growth and welfare. Industry also provides funding to support basic research when there is an incentive to invest. Collecting stakeholder benefits is comparable to a chain reaction—it has to occur before the next one happens.

The logistics that connects the innovation development value chain is not an easy task, but the basis of enhancement of the social and economic value in biobanks and corresponding biotech and health research is an effectual and efficient partnership between the public and private sectors. If all actors involved in the innovation process understand the interplay of their roles between research and translation and commercialization in a series of events, and the series fails, will there at least be more opportunities? Any gaps or weaknesses in the described value chain reduce the chances of regaining profits to society or cause expensive inefficiencies in the innovation development.

Therefore, the function of the RRI theory is to direct the contribution of "responsible" actors in the transformation and commercialization procedure and directly innovation shaping process. RRI must not be an end in itself but facilitate interaction between stakeholders to translate and commercialize new translations and to help each other maintain mutual trust and respect for public values. MRC Technology, representing the UK Medical Research Council, may be among the institutions that can better maintain public trust and convert and commercialize biobanking science and the findings that based on it [24]. Specialized organizations that want to overcome the breach between open science and open innovation in some areas, like the autonomous transformation organizations mentioned above, cooperate openly with specialized industrial segments to find novel methods to incorporate and use university research to progress with new inventions. With the exclusive understanding of the specific workflows relevant to the field and the participation of each stakeholder group, these objective translation science organizations will essentially act as "curators" and resolve how donor rights are used to accomplish the most anticipated economic results. For instance, the specific way in which open science and intellectual property are dealt with depends on the trust of organizations that believe in granting access to the latest translations and, if necessary, approving and distributing licenses for later inventions.

Different industries and regulatory structures affect each industry sector, and since translation companies operate in a wide range of educational institutions involved in the scientific investigations and improvement actions of a specific industry, these organizations may be in a very unbiassed and informative state to find the best combination for the public. Science, research investigations and intellectual property rights serve to guide the transformation and possible commercialization of reliable results of research and to achieve the most wanted socio-economic outcomes. Biobank donors do not essentially create obstacles or close the door to knowledge, because there is a suitable levelness between the "spectrum between the free use of knowledge by anyone for any purpose and the exclusive use of an organization for its own use". Translation and commercialization clarify the intentions and objectives of how public health benefits are provided, and then people commercialize efforts, asset management, RDI management principles, interrelationships and appropriate management counterpart actors.

References

1. Personalized Medicine Coalition. (2014). *The case for personalized medicine* (4th ed.). Accessed December 15, 2015, from http://www.personalizedmedicinecoalition.org/Userfiles/PMC-Corporate/file/pmc_case_for_personalized_medicine.pdf
2. Parks, A. (2009). *10 ideas changing the world right now.* Accessed December 15, 2015, from http://content.time.com/time/specials/packages/article/0,28804,1884779_1884782_1884766,00.html
3. Davey Smith, G., Ebrahim, S., Lewis, S., Hansell, A. L., Palmer, L. J., & Burton, P. R. (2005). Genetic epidemiology and public health: Hope, hype, and future prospects. *Lancet, 366,* 1484–1498. https://doi.org/10.1016/s0140-6736(05)67601-5
4. Vaught, J. B., Henderson, M. K., & Compton, C. C. (2012). Biospecimens and biorepositoies: From afterthought to science. *Cancer Epidemiology, Biomarkers & Prevention, 21*(2), 253–255. https://doi.org/10.1158/1055-9965.EPI-11-1179
5. Supra note 102, at 300–301.
6. Innovative Medicines Initiative 2, 2009–2014. Accessed December 15, 2015, from http://www.iscintelligence.com/archivos_subidos/5_fatiha_sadallah_imi2-_2009=2014_vf.pdf
7. International Society for Quality in Health Care. (2020). http://www.isqua.org/education/partner-activities/the-office-of-biobank-education-and-research
8. Mora, M., Angelini, C., Bignami, F., Bodin, A. M., Crimi, M., Di Donato, J. H., Felice, A., Jaeger, C., Karcagi, V., LeCam, Y., Lynn, S., Meznaric, M., Moggio, M., Monaco, L., Politano, L., de la Paz, M. P., Saker, S., Schneiderat, P., Ensini, M., et al. (2015). The EuroBioBank network: 10 years of hands-on experience of collaborative, transnational biobanking for rare diseases. *European Journal of Human Genetics, 23*(9), 1116–1123. https://doi.org/10.1038/ejhg.2014.272
9. Andreoli-Versbach, P., & Mueller-Langer, F. (2014). Open access to data: An ideal professed but not practised. *Research Policy, 43*(9), 1621–1633. https://doi.org/10.1016/j.respol.2014.04.008
10. Thursby, M., Thursby, J., Haeussler, C., & Jiang, L. (2009). *Do academic scientists share information with their colleagues?* Not Necessarily. Accessed July 06, 2021, from http://www.voxeu.org/article/why-don-t-academic-scientists-share-information-their-colleagues
11. Ziegler, N., Gassmann, O., & Friesike, S. (2013). Why do firms give away their patents for free? *World Patent Information, 37.* https://doi.org/10.1016/j.wpi.2013.12.002
12. European Commission. (2013). *Innovation: How to convert research into commercial success story?* Part 3: Innovation Management for Practitioners. Accessed July 06, 2021, from http://publications.europa.eu/resource/cellar/14ff6fe1-023d-4291-8c4d-4ae779aa78a4.0001.02/DOC_1
13. Council of Europe, The European Convention on Human Rights and Biomedicine. (1997). Article 21 states '[t]he human body and its parts shall not, as such, give rise to financial gain'. See also UNESCO, Universal Declaration on the Human Genome and Human Rights (1997), Article 4 which states '[t]he human genome in its natural state shall not give rise to financial gains'.

14. Evers, K., Forsberg, J., & Hansson, M. (2012). Commercialization of biobanks. *Biopreservation and Biobanking, 10*(1), 45–47. https://doi.org/10.1089/bio.2011.0041
15. Capron, A. M., Mauron, A., Elger, B. S., Boggio, A., Ganguli-Mitra, A., & Biller-Andorno, N. (2009). Ethical norms and the international governance of genetic databases and biobanks: Findings from an international study. *Kennedy Institute of Ethics Journal, 19*(2), 101–124. https://doi.org/10.1353/ken.0.0278
16. Haddow, G., Laurie, G., Cunningham-Burley, S., & Hunter, K. G. (2007). Tackling community concerns about commercialisation and genetic research: A modest interdisciplinary proposal. *Social Science & Medicine, 64*(2), 272–282. https://doi.org/10.1016/j.socscimed.2006.08.028
17. Birch, K. (2012). Knowledge, place, and power: Geographies of value in the bioeconomy. *New Genetics and Society, 31*(2). https://doi.org/10.1080/14636778.2012.662051
18. Birch, K., & Tyfield, D. (2013). Theorizing the bioeconomy: Biovalue, biocapital, bioeconomics or . . . what? *Science, Technology, & Human Values, 38*(3), 299–327. https://doi.org/10.1177/0162243912442398
19. Turner, A., Dallaire-Fortier, C., & Murtagh, M. J. (2013). Biobank economics and the 'commercialization problem'. *Spontaneous Generations: A Journal for the History and Philosophy of Science, 7*(1), 69–80. https://doi.org/10.4245/sponge.v7i1.19555
20. Einsiedel, E., & Sheremeta, L. (2005). Biobanks and the challenges of commercialization. In C. W. Sensen (Ed.), *Handbook of genome research. Genomics, proteomics, metabolomics, bioinformatics, ethical and legal issues* (pp. 537–559). Wiley-VCH Verlag GmbH & KGaA. https://doi.org/10.1002/9783527619733
21. Critchley, C., Nicol, D., & Otlowski, M. (2015). The impact of commercialisation and genetic data sharing arrangements on public trust and the intention to participate in biobank research. *Public Health Genomics, 18*(3), 160–172. https://doi.org/10.1159/000375441
22. Shore, C., & McLauchlan, L. (2012). 'Third Mission' activities, commercialization and academic entrepreneurs. *European Association of Social Anthropologists, 20*(3). https://doi.org/10.1111/j.1469-8676.2012.00207.x
23. Clarysse, B., Wright, M., Bruneel, J., & Mahajan, A. (2014). Creating value in ecosystems: Crossing the chasm between knowledge and business ecosystems. *Research Policy, Elsevier, 43*(7), 1164–1176. https://doi.org/10.1016/j.respol.2014.04.014
24. Lead Discovery Center. (2021). Accessed July 06, 2021, from http://www.lead-discovery.de/

Running a Biobank Network

17

Svetlana Gramatiuk, Berthold Huppertz, and Karine Sargsyan

Abstract

In the previous chapters, many issues related to the development plan of sustainable and long-term biobank networks in Ukraine were established and successfully resolved. An interesting finding was that there was a complete lack of understanding about the structure of biobanks in Ukraine. There was no regulatory document to establish a biobank. Scientific, medical and pharmaceutical institutes and clinics had already collected, processed and stored biological material, but neither with standards nor with the application of standard operating procedures. This situation led to a large number of errors in various scientific departments, which caused the reduction of competitiveness of medical sciences in the Ukraine in general. This chapter lists the problems and solutions during the implementation process of the biobanks network "Ukraine Association of Biobanks".

The Ukrainian Showcase

S. Gramatiuk
Institute of Cellular Biorehabilitation, Kharkiv, Ukraine

B. Huppertz
Division of Cell Biology, Histology and Embryology, Gottfried Schatz Research Center, Medical University of Graz, Graz, Austria

K. Sargsyan (✉)
International Biobanking and Education, Medical University of Graz, Graz, Austria

Department of Medical Genetics, Yerevan State Medical University, Yerevan, Armenia

Ministry of Health of the Republic of Armenia, Yerevan, Armenia
e-mail: karine.sargsyan@medunigraz.at

Keywords

Ukraine Association of Biobanks · Implementation biobanking network · Sustainability plan · Financial development plan · Quality management · Risk management · Biobanking IT system · Sample management

In the previous chapters, numerous issues associated with the plan to develop sustainable and long-term biobank networks in the Ukraine were successfully resolved.

An interesting perception was that there was a complete lack of understanding of the biobanks' structure in the Ukraine. There were no regulatory documents available needed to create a biobank. Scientific, medical and pharmaceutical institutes and clinics had already collected, processed and stored biological material; however, neither with standards nor with usage of standard operating procedures. This situation led to a huge number of mistakes in various scientific departments, which caused the reduction of the competitiveness of medical sciences in the Ukraine in general.

Another problematic area of the biobank network in the Ukraine was connected to the resellers who were involved in the resale of services related to biological material. Their main focus was on the transfer of tissues and biological samples from primary sources to research groups. Nearly all of the employees of such companies were without medical education. This situation led to even more mistakes.

The progress of medical and pharmaceutical research in the world, as in the Ukraine, directly depends on the quality of human biological samples. The slowdown of development is directly linked to the limited availability of high-quality samples [1]. The material collected by biobanks from numerous patients provides an opportunity to combine the analysis of medical data with the results of molecular and genetic examination, including morbidity and environmental data [2]. Biobanks may differ in their structure and may have a diversified specialization in the collection of tissues, for example, difficult-to-reach tissues are collected—eyes, brain, and bones; others focus on isolating cells from donor material to create donor-specific cell lines.

Biobanks play a leading role in research efforts in fields such as oncology, virology and genetics, giving an opportunity for medical science to continue its development [3]. The research based on biobank samples mostly aims at identifying key pathogenetic mechanisms of disease development, signalling molecules, proteins and genes, which are involved in disease development. The usage of this information may aid in detecting early stages of disease and in the development of new treatment methods.

Pharmaceutical companies use specific high-end diagnostics, molecular analysis, tissue microchips and genetic screening to identify new "biomarkers" associated with the prediction or diagnosis of specific diseases and/or to assess the therapy effectiveness during treatment.

Another challenge faced by biobanks around the world as well as in the Ukraine is the difficulty of calculating the total costs of providing biosafety services and analysing clinical data, biostatistics data and the allocation of sample costs and determining the cost of maintaining bio-samples at an adequate cost. All this needs to be consistent with the market as well as reflect the real costs of biobanking services and specific specimen storage. The development and implementation of an adequate biobank model, based on the compensation for expenses, remains one of the unsolved challenges for a high number of biobanks all over the world.

At the same time, the acceptance of the monetary challenges related to the construction and maintenance of biobanks is crucial for the successful usage of both public and private support, especially in developing countries.

The innovative aspect of the development of a Ukrainian biobank network is the examination of the development of a biosecurity cost model in the Ukraine, which takes the viable financing mechanisms into account (e.g., the cost recovery models) to ensure a long-term sustainability of biobanks.

The goals of this network project—according to the sustainable development of biobanks—were the elaboration and implementation of a comprehensive plan for a sustainable development, including short-term strategies and solutions of urgent financial problems, long-term fiscal and usable sustainability through targeted transaction revelation and optimization of work areas.

In order to achieve these goals, a multidisciplinary review of the biotechnology market was conducted to refine and collect information on economic models. We conducted the screening for pricing in various European and American biobanks, based on online surveys through e-mail distribution (based on the BEMT questionnaire model), according to the economic characteristics of developing countries. The survey consisted of 30 questions divided into 5 scales, each of which had its own subscales. It included demographic data (country where biobank was located) and the structure of a biobank; sources of financing and cost calculation; mechanisms of price formation for samples; services provided by biobanks; and ways of sample quality control stored in the biobank. We sent questionnaires via email to 90 executives and directors. According to the result of our research and analysis of the results, we have decided to phase out the project for evolving a sustainable business plan as well as a model of sustainable financing.

The diversity of collection strategies of biobanks requires mental and practical planning skills in the field of biosafety. To develop a sustainable business plan for biotechnology in Eastern Europe, it is necessary to have a structured approach for planning and strategic development, including all standard and well-known business model sections. In addition, from our survey, we have estimated questions about potential collections to give an opportunity to be applied in practice over a long period, meeting all the needs of pharmaceutical companies and research medical centres.

Thus, the structure of the Ukrainian biobank network included 16 medical hospitals with diverse interests, as well as 2 departments of the Kharkov Medical Academy of postgraduate education. At the last stage of the formation, the clinic that specialized on cellular lines and cultures also joined, which further provided some financial stability.

At present, the network consists of seven leading cancer hospitals in the Ukraine, including large university hospitals. The presence of a children's regional oncology centre in the network is highly important. The Ukraine Association of Biobanks focuses primarily on cancer and metabolic diseases, as their percentage in the country increases every year. Thus, as of today, there are more than 23,000 patients' samples in the biobanks of the Ukraine, which have been processed, aliquoted and stored.

The association has an intense exchange with patients as strong beneficiaries and regularly advises with some groups of lawyers and community organizations of protecting patients' rights. Moreover, researchers, as well as medical and pharmaceutical companies of the Ukraine, are other beneficiaries, where exchange is important.

Financing of the Ukraine Association of Biobanks was modest and limited by private investments, scientific grants from biopharmaceutical companies, as well as commercial and voluntary donations. We hope that scientific projects and the implementation of sampling storage and processing services will help to ensure sustainable financing in the future.

According to the task of solving a multitude of judicial and legal problems, we have made a decision on the creation of the Ukraine Association of Biobanks. The association is a nonprofit association of enterprises, created in agreement with the legalization of the Ukraine. The Association carries out its activities as a self-regulating organization installed for protection of rights and benefits of members of the Association in the Ukraine and abroad, to pursue scientific research and to promote scientific achievements, to ensure interoperability as well as to provide actions in accordance to ethical and legal requirements. In its activities, the Association does not pursue commercial objectives and does not have the purpose of obtaining profit for its subsequent distribution among members of the Association. Participation in the Association does not impose on its members any restrictions on their own economic activity or any other kind of activity.

The Ukraine Association of Biobanks has the right to act in relations with other persons on its own behalf and on behalf of its members. Moreover, the Association has the right to represent interests in relevant state bodies, enterprises, academic institutions and organizations of any configuration of proprietorship and to speak on behalf of their members within the powers granted by its Charter and corresponding decisions of the General Meeting of Association's members. Also, the Association has the right to request information in the organs of state power and corporation of regional particular direction, enterprises, institutions and organizations of any form of ownership and to submit for consideration of the bodies of state power and organizations of local independent sector proposals on the construction and realisation of the relevant state policy and implemented by local governmental authorities [4].

The process of establishing the association moved slowly due to the limited funding and the lack of certain legal frameworks specific to biosafety activity in the Ukraine. The process was mainly supported by the joint efforts of several individual activists on a personal level.

There are also additional challenges, including the harmonization of standard operational procedures, the centralization of the database and the usage of a unified form of informed consent.

After reaching a consensus among the active parties, the Ukraine Association of Biobanks has begun to liaise with research committees, the National Cancer Registry, groups of patient lawyers, researchers and other stakeholders. The association model is based on several international networks and relates to most international biosafety groups—the Marble Arch, ISBER and ESBB International Working Group.

Thus, at the first stage, a strategy and plan for the formation of a bioethics committee and the necessary legal documents were developed.

In the Ukraine during the last 5 years, the centre of scientific research in medicine and biotechnology was the only bioethics organization, which worked under the Cabinet of Ministers of Ukraine, and thanks to this document, the further global collaboration in the field of ethics became possible. In the Ukraine, most ethical commissions are under organizational control of the council of each medical centre and/or hospital. At present, the Committee on Bioethics works at the National Academy of Sciences of Ukraine (Decree No. 288 of November 7, 2007).

In connection with this, one of the main goals of the network project related to bioethics was the creation of condition-coordinated activities of ethical committees at all levels in Ukraine based on international cases [5].

Patients in the Ukraine donated biosamples for specific research groups and were informed about specific issues and objectives of the scientific project.

After the registration of the Ukraine Association of Biobanks, a standard application form throughout the network was introduced, and an independent ethics committee has been created becoming an important part of the network.

The Ukraine Association of Biobanks firstly provided a standard patient and healthy volunteer's information letter, as well as a form of informed consent for affiliates for material taken from post-mortem cases, the form of informed consent with respect to biobank and a standard access policy.

The Irish way was recognized as the most effective for Ukrainian local conditions, and it was decided to implement a similar sequence of development and submission of documents to the Committee on Bioethics. A consensus was reached with each hospital and centre that entered the biobank network.

At the second stage, the departments of risk and legal management conducted a review of documents, and along with an independent lawyer specialized in medical research, a legal status was received. Thirdly, the Ukrainian Association assessed all the documentation within the country. Fourthly, the Data Protection Commissioner evaluated all documents, paying particular attention to the proposed method of data sharing. Fifth, applications were acquired to the study and ethics committees of each clinical partner in the biobank network.

This way of approval of the documentation has been done for all standard operating procedures used within the network of biobanks.

From the beginning, the association used the SOPs, recommended by ESBB for tissue collection, handling and storage. The SOPs, which are currently used in the entire Ukrainian biotechnology network, are based on the ESBB templates on these procedures with insignificant changes and customization for local requirements. The sensible, reliable and high-grade quality conscious implementation of SOPs within the framework of associations and in compliance with new ISO/DIS 20387 standards will result in efficient cooperation within biobanks and in biomarker research. This way standardization is accomplished through workshops and education on data preservation, database management system government, data interchange, workmanship for collection and storage, ethical considerations, accessibility to specimens and quality control [6].

With the creation of the Ukraine Association of Biobanks, the sequence for the development, revision and approval of documents and SOPs was approved. It is very important that the policies of this network are consistent with national and European decrees, principles and instructions.

We looked through numerous international SAPs [6–9]. The development of a general SAP required very deliberate and delicate negotiations within the network. The differentiation of the factors of acceptance for access to samples and the priority research projects for Ukrainian research groups was long, but decisive. The agreed SAP is a key milestone for a well-functioning network and provides researchers with access to samples in a simple and transparent way.

The Steering Committee of the Ukraine Association of Biobanks (SCUAB) examines applications for samples from medical research teams and pharmaceutical companies. The representatives from each research group in the association are appointed to the SCUAB commission.

The SCUAB is a group of research professionals in the field of biosafety, who recognize the international, local and national significance of biobanks.

In an ideal biobank project, the collected samples are gathered and stored taking into the account future developments in the field of molecular and genetic research as well as strategic directions for the development of the medical sector.

It would not be economically feasible, if in a few years it becomes obvious that already collected biological samples do not possess the necessary characteristics and are not suitable for further research. Considering the significance of this potential, we decided that efforts will be focused on sampling according to the latest innovations.

Therefore, the result of the network's strategy was the decision and organization of high-quality collections and transfer of samples with preservation of the storage temperature to biobanks of the association (ASK-Health Biobank). The ASK-Health Interaction Team Biobank works with every site that collects samples through the SMART IT system.

The SMART IT system monitors and provides information on the availability of relevant supplies on clinical sites and that all samples, sent to a biobank for processing, are previously registered and monitored for providing the safe and timely delivery of all clinical collections.

The above principles and systems apply to sample management after completion of processing: sample transportation, clinical collection, sample registration in the database, sample research, communication with clients via e-mail or internet systems, distribution of biological samples, quality control, preparation of samples and storage. The special system has also been developed for the sustainable development of biobanks, providing an access to information for all types of samples and their quantity, which provides a direct interaction with the project managers.

Risk management is part of all business operations. In biobanks, risk reduction is a very important part of every day's activities. Reducing the potential risks for storing rare biomaterial/cell lines, clinical data and additional important biobanking components such as cellular preparations and active bio- or pharma components requires exceptionally precise planning. The threats can be sectioned to the subsequent categories: reputational risks, ethical risks, financial risks, operational risks, standard laboratory risks, personnel risks, infrastructure risks, IT risks, strategic risks and financing risks.

As it was previously mentioned, all the samples are collected according to a strict standard procedure. It has been decided that quality control in the Ukrainian biobank network is performed on 100% of the collected samples. Each sample is handled according to the SOPs to guarantee samples of the supreme quality are delivered to the customers. The Ukraine Association of Biobanks is also undergoing a study under the IBBL program for quality assurance of biological samples held by an independent laboratory centre (http://biospecimenpt.ibbl.lu).

Thus, methods have been established for both long and functional quality control of biological samples, as well as for operational memory and verification procedures.

The Ukraine Association of Biobanks is convinced that the real factor for stable financing is employees' participation and employees' development, especially of the staff of the various biotechnology departments and scientific projects.

The scientific activity of the Ukraine Association of Biobanks has led to a leading position in the scientific areas of interest, especially virology and cancer, which are fundamental and need translational research. The interaction between the host's immune system and tumours and neuro-oncology is the leading scientific direction of the biobank network.

Therefore, the scientists from the research team at ASK-Health Biobank study various factors that affect the quality of human specimens. This information is used while updating the SOPs to compile with high-quality human bio-specimens.

The staff of the scientific department of the biobank also works at the grant programs in the field of oncology: screening systems for onco-pathology based on phosphorescence; study of the energy potential of the cells for prediction of the oncological disease course; and piezo-effect of the cells as a mechanism of occurrence of onco-pathology.

These research programs have successfully financed the Ukrainian Association of Biobanks during the last 2 years. In the perspective direction of scientific research, the study of stem cells intended to be used in the treatment of various diseases is in consideration, as well as the development and testing of requirements for the processing and storage of cell cultures.

The Ukrainian Association of Biobanks has been introduced and certified to the new standard ISO 9001: 2015. As a supplier of biomaterials for research on human material, the network is proud to be one of the first in the Ukraine to achieve the specified certification version of the ISO 9001 standard from 2015. The Ukrainian

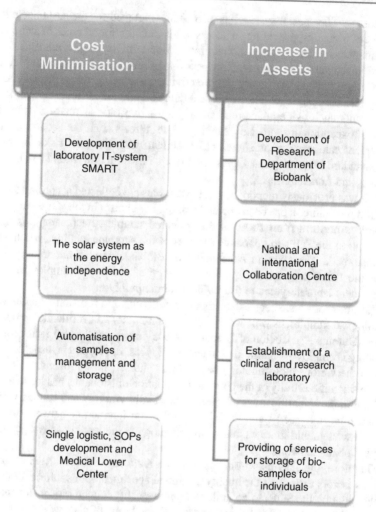

Fig. 17.1 Sustainability plan of the Ukraine Association of Biobanks. © Ukraine Association Biobank

Association of Biobanks was independently evaluated by the experts in this area in December 2016 and has received a certification without any serious inconsistencies from the Bureau Veritas, and in December 2017, they were successfully audited.

Because of this study, a concept for a sustainable financial development plan was developed, which is presented in Fig. 17.1.

References

1. Friede, A., Grossman, R., Hunt, R., Li, R. M., & Stern, S. (Eds.). (2003). *National biospecimen network blueprint*. Constella Group.
2. Botti, G., Franco, R., Cantile, M., Ciliberto, G., & Ascierto, P. A. (2012). Tumor biobanks in translational medicine. *Journal of Translational Medicine, 10*(204). https://doi.org/10.1186/1479-5876-10-204
3. Hanif, Z., Sufiyan, N., Patel, M., & Akhtar, M. Z. (2018). Role of biobanks in transplantation. *Annals of Medicine and Surgery (Lond), 28,* 30–33. https://doi.org/10.1016/j.amsu.2018.02.007
4. Hirtzlin, I., Dubreuil, C., Préaubert, N., & Duchier, J. (2003). An empirical survey on biobanking of human genetic material and data in six EU countries. *European Journal of Human Genetics, 11*(6), 475–488. https://doi.org/10.1038/sj.ejhg.5201007
5. Sándor, J., Bárd, P., Tamburrini, C., & Tännsjö, T. (2012). The case of biobank with the law: Between a legal and scientific fiction. *Journal of Medical Ethics, 38*(6), 347–350. https://doi.org/10.1136/jme.2010.041632
6. Sudlow, C., Gallacher, J., Allen, N., Beral, V., Burton, P., Danesh, J., Downey, P., Elliott, P., Green, J., Landray, M., Liu, B., Matthews, P., Ong, G., Pell, J., Silman, A., Young, A., Sprosen, T., Pearkman, T., & Collins, R. (2015). UK biobank: An open access resource for identifying the causes of a wide range of complex diseases of middle and old age. *PLoS Medicine, 12*(3). https://doi.org/10.1371/journal.pmed.1001779
7. Knoppers, B. M. (2005). Biobanking: International norms. *The Journal of Law. Medicine & Ethics*. https://doi.org/10.1111/j.1748-720X.2005.tb00205.x
8. Vaught, J. B. (2006). Biorepository and biospecimen science: A new focus for CEBP. *Cancer Epidemiology, Biomarkers & Prevention, 15*(9), 1572–1573. https://doi.org/10.1158/1055-9965.EPI-06-0632
9. De Souza, Y. G., & Greenspan, J. S. (2014). Biobanking past, present and future: Responsibilities and benefits. *AIDS, 27*(3), 303–312. https://doi.org/10.1097/QAD.0b013e32835c1244

Infrastructure: Delivering on the Promise of Biobanking

18

Erik Steinfelder

Abstract

Researchers are constantly looking for high-quality samples and associated data stored under the right conditions. Hence, for many it is still difficult to find the right biobanks and to gain access to the collections. This puts the sustainability of individual biobanks at risk. At the same time, the samples and associated data are crucial for biomedical research not only for the development of new drugs and new treatments but also to support developments in prevention. In order to drive research and generally promote innovation as efficiently and effectively as possible, the European Commission has responded with an initiative called Research Infrastructures. Research infrastructures cover many scientific topics such as translational medicine, social sciences and marine biology to advance knowledge and technology. This chapter lists some services that can help biobanks find solutions to their problems in the daily work process.

Keywords

Initiatives · European Commission · Research infrastructure · Services · BBMRI-ERIC · ADOPT project

Researchers are constantly looking for high-quality samples and associated data that are stored under the right conditions [1]. However, for many it is still difficult to find the right biobanks and get access to the collections. In daily routine the complex legal and ethical frameworks in combination with different procedures are creating such hurdles that currently less than 5% of the stored samples in Europe are actually used [2]. Hence, sustainability of individual biobanks runs into mischief. At the

E. Steinfelder (✉)
Thermo Fisher Scientific, Eindhoven, Netherlands
e-mail: erik.steinfelder@thermofisher.com

same time, the samples and associated data are crucial for biomedical research not only for the development of new drugs and new treatments but also to support developments in prevention. This would be impossible without biobanking.

To drive research and foster innovation as efficient and effective as possible in general the European Commission responded with an initiative called Research Infrastructures [3]. Research Infrastructures cover many scientific topics like translational medicine, social sciences and marine biology to advance knowledge and technology. In these new research ecosystems, involving many different countries, researchers have shared access to a variety of services that can support their journey to find solutions for problems that society faces today.

BBMRI-ERIC

On December 3, 2013, a Research Infrastructure, with the special legal status of an ERIC (European Research Infrastructure Consortium), dedicated to biobanking started and was named BBMRI-ERIC (Biobanking and BioMolecular resources Research Infrastructure-ERIC) [4]. Founding member countries were Austria, Belgium, Estonia, Germany, Greece, France, Italy, Malta, Netherlands and Sweden. Observer countries were Norway, Poland, Swiss and Turkey.

Building a biobank community, currently connecting over 500 biobanks, and having a joined approach on quality, IT and ELSI were the cornerstones of the activities of BBMRI-ERIC. Since the participating member states also created an infrastructure on a national level ("BBMRI.xx program"), a distributed European approach was possible creating an overarching structure with common goals but also flexibility to address country-specific challenges.

Knowledge Hub of BBMRI-ERIC

The quality activities in the so-called knowledge hub brought together over 100 biobank quality experts who can be consulted for questions and guidance in international standards that are relevant for biobanking and biomedical research. An instrumental contribution was delivered to the creation of the new ISO 20387 biobank standard, launched in October 2018 [5]. The same group of experts continued educating the wider audience on compliance, the advantages, challenges and opportunities when looking at the details of quality. Biobanks can also assess their own internal processes via the BBMRI-ERIC Self-Assessment Survey or peer-reviewed style audits on request [6].

ELSI Group of BBMRI-ERIC

The ELSI group was instrumental in creating the self-service knowledge base and federated ELSI helpdesk, where support on topics like access policy and MTA/DTA (material transfer agreement/data transfer agreement) can be given via a network of experts. In response to the launch of the EU General Data Protection Regulation (GDPR) on May 2018 an initiative for a code of conduct on Health Research started, providing guidance to researchers and administrative staff, reducing unnecessary fear related to compliance and enhancing data sharing [7].

IT Group of BBMRI-ERIC

Samples without associated data have little to no value for the majority of researchers, and therefore IT-related services were on the program from day 1. Several tools were launched to support researchers in finding material, also enabling effective communication between the parties involved. Services for newly established biobanks or biobanks lacking sufficient IT systems are also offered, and in the near future, data harmonization services are planned. BBMRI-ERIC leads the development of provenance information standard in the ISO Working Group TC 276 in order to allow computer-based assessment of the quality of samples and data.

Sample Directory of BBMRI-ERIC

Improving access is a key indicator to measure the success of BBMRI-ERIC in general where the described services should enable this on a more detailed level to the various stakeholders. The Directory, for example, a catalogue containing collections from biobanks all over Europe, creates the possibility to simply browse biobanks by name or search sample collections according to different criteria such as kind of material, diagnosis, country, etc. [8]. Since it allows researchers to check if there are samples matching their criteria, the Directory is a very powerful tool, especially if someone is in the early stage of writing a project proposal or developing a research project. There are great collections and in Europe, over 100 million samples are stored but as mentioned hardly 5% are used on average. Monitoring the number of visits, requests or samples or cooperation and successful matches on a monthly basis is a good indicator to see if access is really increasing.

The ADOPT Project and Beyond

Executing on the improving access strategy and achieving the desired goals require funding and resources. Research and Innovation Actions (RIA) were activities within Horizon 2020 that aimed for establishing new knowledge and/or to explore the feasibility of new or improved technologies, services or solutions, supported with in most cases 100% funding of the activities. A Research Infrastructure could apply for this type of grant and BBMRI-ERIC was successful with the ADOPT proposal (implementAtion anD OPeration of the gateway for healTh into BBMRI-ERIC). The ADOPT project started in 2015 and aimed at boosting accelerating the implementation of BBMRI-ERIC and its services [9]. Aware of the challenges in getting access and sharing data across Europe and the desire from the funders to increase the actual use of the collections, a dedicated task was created to map the challenges to be overcome and develop a blue print on how to share data sets from patients across Europe, who were in this specific case diagnosed with colorectal cancer (CRC). To achieve this, a common data model was defined, a data protection policy written, the recruitment process defined and how participating biobanks could be reimbursed. During a period of 30 months, the project was executed and delivered an impressive result; in the end, 25 biobanks from 12 different European countries were able to bring 10,480 cases together [10, 11].

Important learning point is the need of expertise on the ethical and legal side that both have to work constructively with IT to create the needed contracts. An important aspect that shouldn't be underestimated when respecting both national legislation and university policies. A major other learning point in the project was the discussion around reimbursement. There was funding to support the biobanks that delivered the datasets, but in hindsight, this was far from being enough to compensate the real costs involved.

Overall benefit to the wider community is that the models that were developed are now publicly available and can be used for projects and consortia that do want to access pan European datasets. For an individual university to have the bandwidth to achieve something similar is close to impossible and is a waste of valuable resources to reinvent the wheel all the time in every biobank for every contract or sample sharing activity. Projecting this information to, for instance, Ukraine, where colorectal cancer in 2020 is the third cause of cancer death for both men and women, it can significantly advance the research and re-use of data, which is another important aspect of the fact why we are biobanking [12].

Biobanks have a variety of stakeholders, (researchers, clinicians, policymakers, funders, patients, regulatory bodies, industry) that have power, legitimate interest and are willing to take action. If you want to have maximum impact speaking in one voice with a clear mandate, this is the only way forward. Within an individual institution, you can negotiate and focus on specific items relevant for you, but if this is too much of a difference, you can end up with many different contracts and in daily practice this is not workable. This also was a learning point in the described CRC cohort. Within a research institution (RI), you can try to work out consensus and with that agreement you can go to other negotiations, and this is working bi-directional. For policymakers or patient organizations, for instance, it is also more efficient and creates more impact if they can discuss with a network organization or infrastructure versus several separate universities.

When resources and funding are scarce, you have to do it right. This is certainly an obligation to the patients whom you are taking into account. You are actually using their material and data to really foster research and drive innovation. In recent discussions around the GDPR implementation experiences, it was clear that it sometimes creates so many hurdles to actually use the samples in other institutions that they are simply not willing to cooperate or share, afraid of the legal consequences. Research Infrastructures can address this to the right policymakers and have a strong united voice.

Joint Forces of Research Infrastructures

The wish from ESFRI (European Strategy Forum on Research Infrastructures) and BBMRI to actually start measuring KPI's (key performance indicators) should also give more transparency of the results that have been made or where a change in strategy is needed to gain more impact. By doing, this on a national level you can also see where additional help is needed. An extra step that is currently made to have more impact is the fact that in some countries different medical RI's joint forces; Health RI in the Netherlands and AMRI (Alliance of Medical Research Infrastructures) as an initiative on a European scale; here BBMRI-ERIC, EATRIS-ERIC (European Advanced Translational Research Infrastructure in Medicine-ERIC) and ECRIN-ERIC (European Clinical Research Infrastructure Network-ERIC) are joining forces to tackle pressing issues in medicine development for the benefit of the patients. Increasing the visibility and getting access to the stored samples is key, not only in the countries that were there from the beginning, but maybe even more for those biobanks in LMICs (Low- and Middle-Income Countries).

In quite some Western European countries, a strong network of scientists and departments in academic centres already exists and via H2020 scientific excellence was achieved. But how does this work for Eastern Europe that in some cases is still building and not ready for excellence? Another good example where a research infrastructure can help: learn from the ones that have done it already, but also learn from their mistakes. Transparency can help drive research forward and deliver on the promises that biobanks have already for quite some time. Bulgaria joined BBMRI-ERIC in 2019 and Lithuania in 2020, which shows interest from Eastern Europe and connections with the biobank community. In response, one of the activities of BBMRI-ERIC is to develop a strategy for National Nodes to use European Structural and Investment Funds and other capability building opportunities, also supported by Latvia, Poland, Lithuania and Estonia.

Increasing access of samples cannot be achieved without investing in the visibility of the biobank community and the collections that are available for research. Communication is key: What can you do as individual biobank in promoting your own collection? There are of course the networks between universities in one country and beyond and this is happening already within consortia. However, outreach activities only show that average researchers simply are not aware of the availability of catalogues or sample collections in the medium and small size academic institutes. Visibility might even be more important for those that are in LMICs. They manage in many cases unique collections that are of high interest to other academic institutes and biotech companies. Initiatives like B3Africa, the ESBB Africa Working group, and the BBMRI-ERIC ADOPT internationalization project task in the Middle East and Latin America all can help support multiple biobanks in one communication campaign to reach a wider audience.

References

1. Thank you for sharing. (2020). *Nature biotechnology, 38*(9), 1005. https://doi.org/10.1038/s41587-020-0678-x
2. BBMRI-ERIC. *ADOPT Grant Agreement NO. 676550.* Deliverable Report D2.6. (2019). Retrieved July 06, 2021, from https://www.bbmri-eric.eu/wp-content/uploads/D2.6_rev.pdf
3. European Commission. *European Research Infrastructure: What Research Infrastructures are, what the Commission is doing, strategy areas, funding and news.* Retrieved July 06, 2021, from https://ec.europa.eu/info/research-and-innovation/strategy/strategy-2020-2024/our-digital-future/european-research-infrastructures_en
4. BBMRI-ERIC. *Statutes of the biobanking and biomolecular resources research infrastructures. Rev2.* (2016). Retrieved July 06, 2021, from https://www.bbmri-eric.eu/wp-content/uploads/2016/12/BBMRI-ERIC_Statutes_Rev2_for_website.pdf
5. ISO. *ISO 20387:2018.* (2018). Retrieved July 06, 2021, from https://www.iso.org/standard/67888.html
6. BBMRI-ERIC. *BBMRI-ERIC Quality Management Services for basic and applied research.* Retrieved July 06, 2021, from https://www.bbmri-eric.eu/services/quality-management/
7. A Code of Conduct for Health Research. Retrieved July 06, 2021, from http://code-of-conduct-for-health-research.eu/
8. BBMRI-ERIC. *Directory.* Retrieved July 06, 2021, from https://directory.bbmri-eric.eu/
9. BBRMI-ERIC. *Documents, Publications, Media & More.* Retrieved July 06, 2021, from https://www.bbmri-eric.eu/publications/
10. BBMRI-ERIC. *ADOPT Grant Agreement NO. 676550.* Deliverable Report D2.4. (2019). Retrieved July 06, 2021, from https://www.bbmri-eric.eu/wp-content/uploads/D2.4_rev3.pdf
11. BBMRI-ERIC. *ADOPT Grant Agreement NO. 676550.* Deliverable Report D2.7. (2019). Retrieved July 06, 2021, from https://www.bbmri-eric.eu/wp-content/uploads/ADOPT_D2.7_revised.pdf
12. International Agency for Research on Cancer World Health Organization. *Cancer Today.* Retrieved July 06, 2021, from https://gco.iarc.fr/today/online-analysis-map?v=2020&mode=population&mode_population=continents&population=900&populations=900&key=asr&sex=0&cancer=39&type=0&statistic=5&prevalence=0&population_group=0&ages_group%5B%5D=0&ages_group%5B%5D=17&nb_items=10&group_cancer=1&include_nmsc=1&include_nmsc_other=1&projection=natural-earth&color_palette=default&map_scale=quantile&map_nb_colors=5&continent=0&show_ranking=0&rotate=%255B10%252C0%255D

Design Considerations and Equipment of Biobanks

19

Erik Steinfelder

Abstract

Talented researchers, scientists, data specialists and laboratory staff, for example, are without question a fundamental resource for a biobank. Supported by the right tools, they can work on the challenging scientific questions that need answers to better understand certain diseases or to be able to treat them with new drugs in the future. This chapter provides an overview of the main instruments, consumables and data management platforms for biobanks, including some considerations when designing a biobank from scratch.

Keywords

Design of biobank equipment · Key instruments · Consumables · Data management platforms · Automation systems · Maintenance · LIMS/BIMS

Equipment (and Some Design Considerations)

Talented researchers, scientists, data specialists and, for example, lab staff are without question a fundamental resource for a biobank. Supported by the right tools, they can work on the challenging scientific questions that need an answer in order to understand better specific diseases or how to treat them in the future with new drugs. What are the key instruments, consumables and data management platforms to consider? Here is an overview, including some considerations when designing a biobank from scratch.

E. Steinfelder (✉)
Thermo Fisher Scientific, Eindhoven, Netherlands
e-mail: erik.steinfelder@thermofisher.com

Building and Rooms

Multiple universities across the globe have allocated the biobank in the basement of one of their buildings; this seems a logical choice to keep the costs down and with limited space sometimes the only option available. Location selection should, however, not be underestimated, and an assessment of the risks involved for the specific site on, for example, lightning, earthquakes or flooding needs to be made. Facilities should also be prepared to address situations like power failure and emergency storage. Active planning and preparing for risk mitigation, redundancy and monitoring can help.

Consumables and Software

Careful planning and mapping the risks is also needed on the equipment side. Instruments, consumables and software all should be fit for purpose, and therefore it is crucial to have a clear vision on why you are collecting a specific set of samples and which potential research questions they could help to answer. This more strategic approach can also support in finding the right tactics as well as making the correct operational choices that need to be made when preparing for example the needed instruments and their detailed technical specifications.

Collection of Samples

A lot of work already starts outside the biobank site; samples need to be collected, forms to be filled and staff trained to support all the efforts. In this part of the process, there is not always the opportunity to select the samples, collection tubes or 1D barcode characteristics; however, it is crucial to be aware of what is used for which part of the collection process and what to expect once it arrives at the reception desk of the facility. Buy in from other parts of the organization that are crucial for a biobank to be successful might be more important than dictating a specific new tool or product.

Arrival of Samples at Biobank Site

Upon arrival of samples at the biobank site, the correct registration of the samples takes place, and already there some critical time steps will start, and potentially specific actions need to take place. In many cases, the so-called mother sample will receive a new unique ID, also to be compliant with internal anonymization or pseudonymization procedures. Warning flags should be automatically raised when inconstancies occur like missing data, incorrect data, damaged collection tubes or incorrect minimum volume of the material. Depending on internal procedures, this might also be the moment to record if the donor or patient wants to be informed in case of incidental findings.

Automation and Maintenance

Numerous biobanks nowadays are using full or almost full process automation and are taking benefit of the modern well-developed instruments for biospecimen management. Data management tools are considered to streamline instrument and automated system integration. Formerly incongruent biobanking infrastructure components can now be combined to open the possibility for data to be electronically distributed among them, excluding manual error-prone procedures and rising efficiency and data consistency. Instruments itself contain predictive maintenance programs and allow remote diagnostics and performance management to lower the total cost of ownership. Implementing ISO 17025 and/or ISO 20387 (the new biobank standard) can support and capture in these instruments' requirements. According to these ISO standards, equipment should be provided by selected suppliers that are regularly audited. Maintenance should be scheduled, and only properly trained staff is allowed to do specific maintenance, calibration or servicing. Building a track record of when which instrument is checked by who helps to be in control, be ready for audits and increase the quality of operations in general.

Aliquoting and Collection Tubes

Next steps in the process of a unique biobank sample strongly depend on the setup of the biobank and which goals need to be addressed. Assuming several copies will be made from the original sample aliquoting is needed. Here another paramount decision needs to be made first: in which sample storage tube will the aliquot be transferred to? Once the number and volume of the aliquots is decided, a sample storage 2D collection tube can be selected, which is also an important factor in the path towards standardization. The permanent 2D barcode creates a solid, proven way of tracking the sample during its lifetime in the biobank. In combination with barcode readers, sample data can be integrated into existing databases or tracking systems. The tubes will be ready to be utilized for storage in nearly any laboratory condition, from room temperature to the vapour phase of liquid nitrogen.

Choice of Collection Tube

Where there are literally several dozens of different tubes available, biobanks still have too often the tendency to request for a custom-made special tube that would fit their processes better. This is not only challenging for the manufacturer from a quality, stock and logistics perspective but also interoperability becomes an issue. Biobanks providing samples in formats that cannot be handled outside the organization create an additional hurdle for them to be used. Several automated storage systems are modified as well to fit the custom tube, involving additional investment without clear evidence of additional value to research. Supporting instruments like Capper/Decapper can support for optimization of benchtop processing of screw top storage tubes, maintaining crucial sample integrity.

Liquid Handling

Assuming a significant throughput of samples is required, automated liquid handling comes into place. The majority of the solutions provided is able to interface with the Laboratory Information Management System (LIMS) to receive worksheets on which sample to take, how much volume from and in which rack with tubes this should go to. In combination with the 2D barcode scanner and the liquid handler software, the exact location and information of each sample is available, again supporting a full audit trail. From a redundancy point of view, it might be good to invest simultaneously in a backup or second liquid handler, so in case something is not performing as planned, not the whole process stops and the quality of received samples might be in danger.

Cold Sample Storage

Depending on the sample type, the purpose of storing and the period of storage, the temperature conditions can be determined [1]. All matrices have their own optimal storage conditions, and handling at temperatures outside these ranges can cause significant sample stress as a result of cellular dehydration or ice formation. Similar as with other steps in the process, careful planning is crucial to maintain sample integrity and to prevent samples from becoming too warm or too freeze due to direct contact with dry-ice. Here the best option to keep a sample between +1 °C and +10 °C is to use a water/ice slurry. The sample type should also determine the rate at which a sample is cooled for storage and eventual thawing. When selecting the right cold storage solutions, keep in mind the fit for purpose; domestic, residential or white boxes do not meet the rigid temperature control, stability and uniformity requirements.

Backup Storage

Backup storage units that already have the right temperature, have ample reserve of LN_2, dry ice and have generator fuel onsite are critical. Obvious measures, but surprises can still occur and thawing of frozen samples and freezing of refrigerated samples should be prevented. Active and regular monitoring beyond only temperature, but also on relative humidity and/or CO_2 concentration, supports to identify issues at an early stage, or even prevent them from happening altogether. Secure data logging and automatic audit trials support regulatory compliance as mentioned earlier.

Automated Cold Storage

In recent years, the interest in automated −80 °C storage has increased and implemented in a number of biobanks in Europe, Middle East, Asia and North America. Reasons for making the investment has been the capacity, the availability of space and the usage of energy on one hand, on the other hand to fulfil projected requests for samples in a high-quality environment. Where this can be true for some biobanks, the majority does not need these capital-intensive investments to run the biobank in line with the highest-quality standards. If there is funding to make larger investments, it might be worth to consider managing the data as best and secure as possible to be ready for future demand.

Laboratory Information Management Systems (LIMS)

The actual registration and many other process steps require a data management system. Changes in technology and connectivity have influenced data management in the last decade significantly. LIMS solutions are already available for decades for various types of industries with hundreds of thousands of users globally every day. Due to the maturity of the solutions, the experience and knowledge of the providers and active user communities, updates and upgrades are available on a regular basis and strong organizations can help out with maintenance and support. Downside is the relative high investment and running costs in comparison to the functionality that is used.

Biobank Information Management Systems (BIMS)

In response, some software developing companies created a BIMS. A BIMS is a dedicated solution that specifically focuses on the challenges within a biobank, not only sample management on who logged which sample at which moment and managing critical times between taking and freezing but also on the consent status of the samples/donor to make sure it is only used in the right studies and it is flagged for what should be done in case of incidental findings. Going forward, there are still discussions on how to do this, and therefore flexibility can be an option. For both LIMS and BIMS, it should be taken into account that it requires dedicated internal resources for first-line support and can also support change requests. Acknowledging the fact that a sample does not have a lot of value for research if it misses the associated data, the investments needed should be budgeted to prevent incorrect data or unclear data provenance.

Homegrown LIMS/BIMS

Inside academic institutions, it is still popular to apply for specific grants to support a small team of computational scientists/software developers and programmers to develop a homegrown LIMS/BIMS. Great advantage is that it can be made exactly to the wishes of the end-user and no perpetual investments in license and support fees are needed. Where this may seem to be an effective option from a cost perspective point of view, but more and more biobanks do not see this as a sustainable way forward.

This kind of tools will be hard to create with a homegrown solution. In LMICs, it is best to go for a proven online system that can do the job. This is the way that the vendor can help out online, etc. If people need to be onsite all the time, you lose too much money on travel and manpower. Keep an eye on security.

Just like biobanking is a discipline on its own requiring specific expertise, so is data management. The often-made remark that data is as important as the sample makes this clear. Invest properly in a data management solution, supported by the right staff. If there is no real IT support available within the biobanking itself, it will be hard to manage.

Reference

1. Hubel, A., Spindler, R., & Skubitz, A. P. (2014). Storage of human biospecimens: Selection of the optimal storage temperature. *Biopreservation and Biobanking, 12*(3), 165–175. https://doi. org/10.1089/bio.2013.0084

Software Tools for Biobanking in LMICs

20

Dominique Anderson, Hocine Bendou, Bettina Kipperer, Kurt Zatloukal, Heimo Müller, and Alan Christoffels

Abstract

The appreciation of the scientific value of biospecimens for current and future basic and translational research is increasingly being recognised. While biobanks in high-income settings are well established, the number of facilities in low- and middle-income countries is growing, albeit at a slower pace. There exist several challenges for biological research centres, which operate in low resource settings, and these need to be taken into consideration when seeking to apply consensus in the biobanking discipline. The overall cost of custodianship should be taken into account, so as not to create, and indirectly promote, exclusivity, exploitation and knowledge silos. Therefore, standards and tools for biospecimen quality management must be democratised for biorepositories in a variety of settings to have a truly global impact on research.

Keywords

Biobanks · Laboratory information management systems · Biospecimen archive · Baobab LIMS™

D. Anderson · H. Bendou · A. Christoffels
South African Medical Research Council Bioinformatics Unit, South African National Bioinformatics Institute, University of the Western Cape, Cape Town, South Africa

B. Kipperer · K. Zatloukal · H. Müller (✉)
Diagnostic and Research Center for Molecular BioMedicine, Diagnostic and Research Institute of Pathology, Medical University of Graz, Graz, Austria
e-mail: heimo.mueller@medunigraz.at

© The Author(s), under exclusive license to Springer Nature Switzerland AG 2022
K. Sargsyan et al. (eds.), *Biobanks in Low- and Middle-Income Countries: Relevance, Setup and Management*, https://doi.org/10.1007/978-3-030-87637-1_20

Introduction

Biological specimens such as cells, body fluids, tissues, plants, seeds, microbes or isolated biomolecules are essential materials for research and development in medicine, agriculture and biotechnology. Specimens, together with associated data are handled by biobanks and/or resource centres in an institutional and well-organised way. The core activities of a biobank or resource centre includes the collection, processing, preservation, storage and custodianship of specimens, together with the associated metadata, and most importantly, the provision of access to these resources [1]. Biobanks support research in a variety of fields, such as agricultural research, human health, rare diseases [2] and biomarker development [3], to mention a few. The electronic cataloguing of samples, workflows and operating procedures of biobanks are supported by several software tools [4], and the Biobank Information Management System, can be considered as a central component, with open-source solutions clearly offering substantial benefits [5].

Biobanking in Africa

Biobanks are repositories, which receive, process, store and disseminate specimens. Biorepositories may be in the form of small, university-based personal research collections, up to large collaborative multi-centre facilities. Well-curated and annotated, high-quality biospecimens which can be provided by these biological resource centres are the cornerstone of biological science, playing a central role in basic, translational and clinical research [6–8]. As such, there is little doubt about the value of biobanking in driving reproducible and trustworthy research.

Notwithstanding biospecimen collections across Africa by individual researchers, the first national databank for DNA was established in the Gambia in 2000 by the UK MRC. This DNA bank contains over 57,000 samples collected from participants in West Africa [9]. Early initiatives in biorepository science on the African continent were subsequently introduced to focus on disease- or population-specific research, with substancial focus on the collection of samples for HIV/AIDS, tuberculosis and malaria research [6–10].

Efforts to strengthen existing collaborative networks between the global north and south, as well as promote and support valuable south-south networks, by building platforms which support African researchers in establishing large interdisciplinary collaborations, is being driven by a multitude of associations [11]. The formation of regional and national bioresource centres in Africa by global consortia has bolstered biorepository science and increased awareness in biobanking, biosecurity, genomics and bioinformatics [12]. For example, consortia such as H3Africa [13], B3Africa [14] BCNet and GetAfrica [15] have catalysed biobanking science on the African continent. Through these initiatives and others, African researchers are delivering innovative, world-class scientific research which is mapping a pathway to health and data ownership on the continent while remaining cognisant of the ethical implications of biospecimen and data custodianship [9, 16, 17].

Biobank QC and Informatics

The drive to bolster biorepository science capacity has been demonstrated by substantial efforts to obtain consensus toward best practices and standards by several agencies within the scientific community, such as ISBER and ESBB. More recently, an ISO standard was developed (ISO 20387: 2018 "Biobanking—General requirements for biobanking"), which documents internationally agreed common operational guidelines and requirements for intercontinental accreditation of biobanks. While this is a welcome development, it is important to note that for biorepositories operating in low- and middle-income countries, the resource cost associated with such a global accreditation would likely be out of the reach of many.

LIMS are integral components of the informatics infrastructure that support biobanking activities and form a valuable part of the quality management roadmap. To improve overall quality of samples, significant events related to the biospecimen lifecycle must be captured and recorded and includes pre-analytical, analytical, and post-analytical laboratory operations [18, 19]. The implementation of appropriate informatics tools for biobanking enhances good laboratory management practices, and despite the heterogeneity in biobank design and individual approaches to data collection, annotation, and processing, improvements in laboratory efficiency, management and overall quality have been observed when a fit-for-purpose LIMS is integrated into the core business operation.

To date, a wide range of commercial and open-source LIMS are available. The decision to opt for one LIMS over another is often influenced by the needs of the biobank clients and researchers, as well as available financial resources [20]. However, to find a LIMS that incorporates all possible requirements of a biobank is a complicated endeavour [20]. Full LIMS implementation takes a considerable amount of time and the complexity, size, function, resource availability and internal data management policy of a biobank must be considered when evaluating the suitability of the LIMS. Both national and international best practices and guidelines should be considered when evaluating a fit-for-purpose LIMS, and due to the overall cost of ownership, this is of particular importance in low resource settings. It is imperative to understand the limitations of a system. Functionality gaps can have a significant impact on end-user perceptions, and any "inefficiencies" may likely result in users returning to previous working methodologies, in order to accomplish tasks [21–23].

Baobab LIMS™: An Open-Source Laboratory Information Management System

The need to implement biobank standard operating procedures as well as the use of standards for biobank data representation was the motivation for the development of an open-source LIMS for biobanking. The decision to incorporate an open-source LIMS, which is robust, easy to use and customisable, was undertaken by the B3Africa consortium, allowing researchers to track the lifecycle of a biospecimen in the laboratory from collection, receipt, storage, analysis and reuse, all while ensuring that sufficient sample metadata is captured (www.baobablims.org). Baobab LIMS™ was developed following an evaluation by stakeholders, and with the aim to automate state-of-the-art biobank operations. One of the motivations for developing this tool was to address the lack of open-source LIMS, which integrate maximum functionalities required by modern biobanks. The existence of such a biobanking tool in the market, which is rich in functionality, well designed and free of license fees, is of great value for biological resource centres, particularly those which operate under limited resources in low- and middle-income countries, specifically on the African continent. Baobab LIMS™ is based on the Plone web-content management framework, a server-client-based system, whereby the end user is able to access the system securely through the internet on a standard web browser, thereby eliminating the need for standalone installations on all machines. The system features were designed *a priori* using SOPs and workflow management of the NHLS-Stellenbosch University Biobank (Fig. 20.1).

Baobab LIMS™ implementation has three phases: installation, data setup and configuration, and workflow operation. The setup phase provides the end-user with a high degree of flexibility to configure the base inputs of the system, which are applicable to their operations. For example, the configuration of the laboratory allows the administrator to specify details of the biobank, such as the address, and accreditation. Furthermore, end users of the system and their designated roles are created in this module and are provided with secure login credentials. Setup

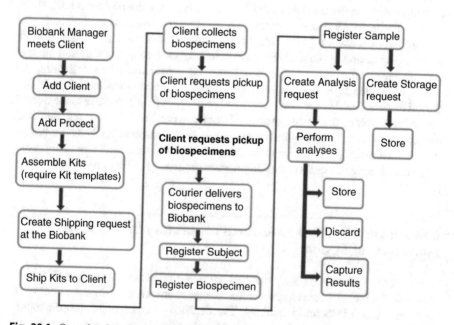

Fig. 20.1 Operational workflow of the NHLS-Stellenbosch University Biobank used to develop and implement Baobab LIMS™

configuration by the biobank is a once-off operation; however, new data can be added as operations of the laboratory expand. Clients, projects, biospecimen kits assembly and shipment, biosample registration, collection and storage, as well as analysis and results capture, are the core functions in the workflow operation phase. [More detailed information can be found in [20] or in the user documentation (www. baobablims.org).]

Enhancements and Continued Development of Baobab LIMS™

Key to continuous development and improvement of Baobab LIMS™ following the B3Africa funding cycle, the team engaged with potential end-users to gain greater insight into the overall heterogeneity of operations in different laboratory and biobank environments. This provided an understanding of the LIMS functionality gaps and promoted development of practical solutions to fit local needs. While the majority of features within Baobab LIMS™ meet requirements for multiple end users, some valuable improvements have been added and are summarised below.

Sample Shipment

To enhance tracking events within the biospecimen life cycle, Baobab LIMS™ developed a sample shipment module, which operates independently from the kit shipment module (see Fig. 20.2a). This functionality is valuable for biobanks, such as the Makerere University biobank in Uganda, which transport samples or sample aliquots, either to internal departments or to external clients, as a core operational process.

High-Throughput Capability

The need to increase the capacity of Baobab LIMS™ for high-throughput capability was an important evolution in the system. The batch sample module (see Fig. 20.2b) was developed for the addition of multiple samples at once, and the design relies on state changes, allowing for the pre-registration of biospecimens prior to receipt.

Meta-data Aware Functionality

Finding the volume and/or diversity of samples, which will add statistical rigor to a study, can be challenging, especially in the case of rare diseases [21, 24]. Biobanks may opt to generate a catalogue of available samples within their facility, making this resource discoverable for the greater research community. BBMRI-ERIC developed MIABIS, a data standard used to detail the minimum amount of information related to a sample, which should be provided for sample shareability [24, 25]. To

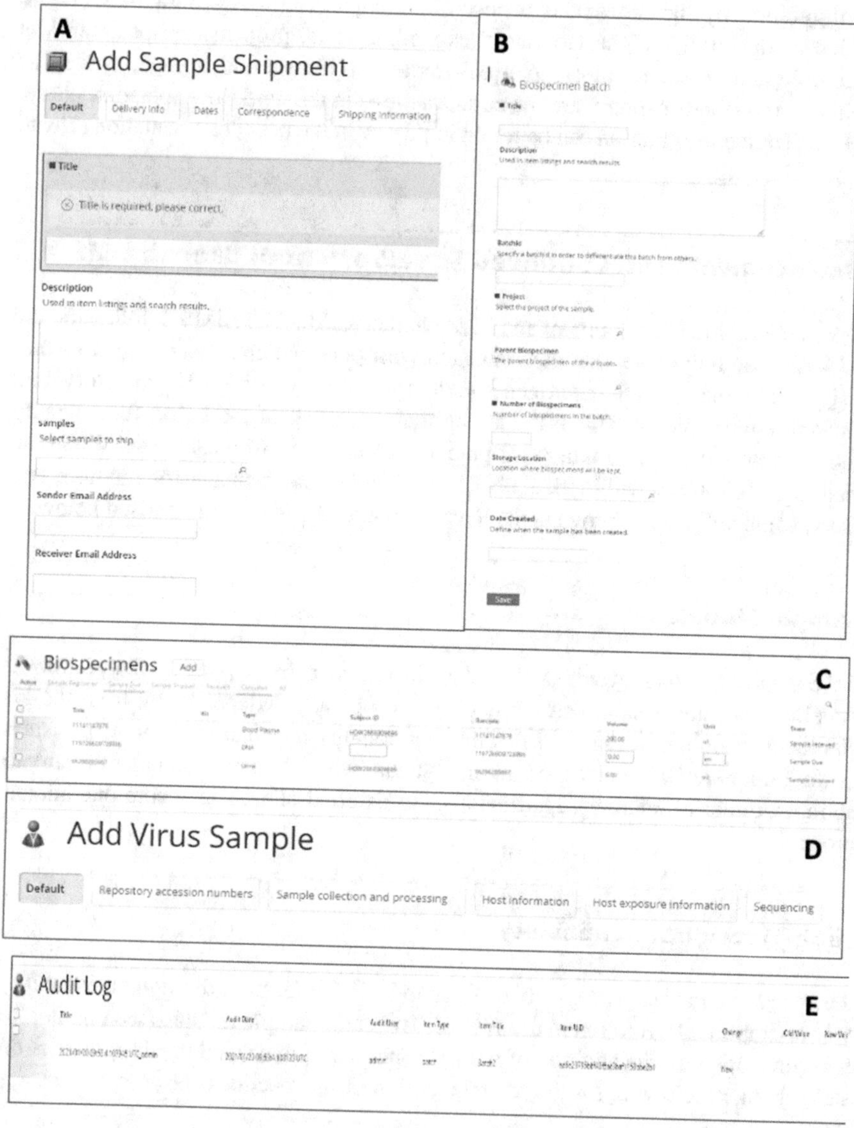

Fig. 20.2 Screenshots of new Baobab LIMS™ features as of January 2021. (**a**) Sample shipment module to enhance tracking events within the biospecimen life cycle; (**b–c**) batch module to improve throughput of sample collection and storage; (**d**) adding virus biospecimens with standardized metadata developed by PHA4GE; (**e**) audit log module for internal quality control

facilitate sample sharing, and data standardisation, Baobab LIMS™ incorporated the MIABIS 2.0 core attributes for sample donors and disease ontology. The approach used by Baobab LIMS™ to allow for configurability while adhering to consensus ontology and format was to build these modules as content types and link data to a specific sample via a drop-down list consisting of prior user input data. This reduces the need for repetitive data entry, thereby limiting errors and data capture fatigue. Incorporation of metadata standards in Baobab LIMS™ is aimed at improving interoperability and harmonisation, and the most recent common structured vocabulary added to Baobab LIMS™ is based on the SARS-COV-2 metadata standard, developed by a working group within the PHA4GE consortium (http://www.pha4ge. org). Using a combination of pre-coded options, as well as configurable content types, Baobab LIMS™ broadened the scope of this module to ensure applicability to the collection of any viral sample (see Fig. 20.2d).

Quality Control and Audit Logs

Additional developments in Baobab LIMS™ include a JSON application-programming interface for programmatic interaction between separate software components and resources to make them more reusable. In terms of a feature gap in Baobab LIMS™ analytics and QC capability, an audit logging functionality was a vital addition to the system (see Fig. 20.2e). Baobab LIMS™ also enhanced the existing Excel file import functionality, as a viable work-around for users with intermittent internet connectivity. The importer includes automatic control checks to ensure that data imported into the system is not duplicated. In line with this, an export functionality has been included in Baobab LIMS™. Other noteworthy enhancements related to end-user support includes a website for access to all documentation as well as an updated demo system for users and a ticket generating help desk.

Software Installation and Verification

As the number of new functionalities in Baobab LIMS™ increased, the need for software verification became a requirement to ensure new code changes did not affect existing functionality, and if development errors occurred, they would be identified early on. The continuous integration tool, Travis Cl, was used for this purpose, and the YAML configuration file specifies automated parsing of set-up data into new Baobab LIMS™ builds. To improve Baobab LIMS™ infrastructure deployment, repositories containing Terraform scripts and an Ansible playbook for deployment of a production ready installation have been made available. Depending on institutional regulations or policies, users can successfully deploy Baobab LIMS™ on local servers, as well as on cloud infrastructure, such as AWS. Baobab LIMS™ recently finalised Plone database replication, which automatically copies all information from the primary database to a secondary database, to prevent any data loss due to failure of the central database and protect the integrity of information.

Future Enhancements and Features

Current developmental work to incorporate freezer monitoring in Baobab LIMS™ is underway. This feature will allow for the direct display of freezer temperature monitoring, captured by data logging instruments. To complement metadata aware capability, Baobab LIMS™ will incorporate standards for bacterial samples in the near future.

Software availability

Website: https://baobablims.org/
Code repository: https://github.com/BaobabLims
Helpdesk email address: help@baobablims.org

Conclusion

There is increasing complexity associated with biospecimen quality and intrinsic features are expanding to include extrinsic features, and a lack of standardised terminology has created a challenge as quality metrics begin to apply to complex biospecimens [26]. In addition, big data is driving large-scale research, and, in the future, biobank facilities will need to support multi-omics projects, which will cross the boundaries of scientific disciplines [27]. To sustain digital transformation in the biobanking field, Baobab LIMS™ will seek to continuously evolve and cater for these needs.

Acknowledgements This work was supported by the European Union's Horizon 2020 research and innovation projects B3Africa (GA No 654404) CY-Biobank (GA No 857122) and HEAP (GA No 874662).

References

1. Müller, H., Dagher, G., Loibner, M., Stumptner, C., Kungl, P., & Zatloukal, K. (2020) Biobanks for life sciences and personalized medicine: Importance of standardization, biosafety, biosecurity, and data management. *Current Opinion in Biotechnology, 65*, 45–51. https://doi.org/10.1016/j.copbio.2019.12.004.
2. Gainotti, S., Torreri, P., Wang, C. M., Reihs, R., Mueller, H., Heslop, E., Roos, M., Badowska, D. M., de Paulis, F., Kodra, Y., Carta, C., Lopez Martín, E., Rangel Miller, V., Filocamo, M., Mora, M., Thompson, M., Rubinstein, Y., Posada de la Paz, M., Monaco, L., & Taruscio, D. (2018). The RD-connect registry & biobank finder: A tool for sharing aggregated data and metadata among rare disease researchers. *European Journal of Human Genetics, 26*(5), 631–643. https://doi.org/10.1038/s41431-017-0085-z
3. Zatloukal, K., Stumptner, C., Kungl, P., & Müller, H. (2018). Biobanks in personalized medicine, expert. *Review of Precision Medicine and Drug Development, 3*(4), 265–273. https://doi.org/10.1080/23808993.2018.1493921
4. Müller, H., et al. (2015). State-of-the-art and future challenges in the integration of biobank catalogues. In A. Holzinger, C. Röcker, M. Ziefle (Eds.), Smart health. Lecture notes in computer science (Vol. 8700). Cham: Springer. https://doi.org/10.1007/978-3-319-16226-3_11

5. Müller, H., Malservet, N., Quinlan, P., Reihs, R., Penicaud, M., Chami, A., Zatloukal, K., & Dagher, G. (2017). From the evaluation of existing solutions to an all-inclusive package for biobanks. *Health and Technology, 7*(1), 89–95. https://doi.org/10.1007/s12553-016-0175-x

6. Gasmelseed, N., Elsir, A. A., DeBlasio, P., & Biunno, I. (2012). Sub-Saharan centralized biorepository for genetic and genomic research. *Science of the Total Environment, 423,* 210–213. https://doi.org/10.1016/j.scitotenv.2010.07.054

7. Vaught, J. (2016). Biobanking and biosecurity initiatives in Africa. *Biopreservation and Biobanking, 14*(5), 355–356. https://doi.org/10.1089/bio.2016.29009.jjv

8. Soo, C. C., Mukomana, F., Hazelhurst, S., & Ramsay, M. (2017). Establishing an academic biobank in a resource-challenged environment. *South African Medical Journal, 107*(6), 486–492. https://doi.org/10.7196/SAMJ.2017.v107i6.12099

9. Mendy, M., Caboux, E., Sylla, B., Dillner, J., Chinquee, J., & Wild, C. (2014). BCNet survey participants. Infrastructure and facilities for human biobanking in low- and middle-income countries: A situation analysis. *Pathobiology, 81,* 252–260. https://doi.org/10.1159/000362093

10. Mayne, E. S., Croxton, T., Abimiku, A., Joloba, M., Kyobe, S., Beiswanger, C. M., Wideroff, L., Guyer, M., Troyer, J., & Kader, M. (2017). Genes for life: Biobanking for genetic research in Africa. *Biopreservation and Biobanking, 15*(2), 93–94. https://doi.org/10.1089/bio.2017.0007

11. Munung, N. S., Mayosi, B. M., & de Vries, J. (2017). Equity in international health research collaborations in Africa: Perceptions and expectations of African researchers. *PLoS One, 12*(10), e0186237. https://doi.org/10.1371/journal.pone.0186237

12. Akinyemi, R. O., Akinwande, K., Diala, S., Adeleye, O., Ajose, A., Issa, K., Owusu, D., Boamah, I., Yahaya, I. S., Jomoh, A. O., Imoh, L., Fakunle, G., Akpalu, A., Sarfo, F., Wahab, K., Sanya, E., Owolabi, L., Obiako, R., Osaigbovo, G., et al. (2018). Biobanking in a challenging African environment: Unique experience from the SIREN project. *Biopreservation and Biobanking, 16*(3), 217–232. https://doi.org/10.1089/bio.2017.0113

13. H3Africa Consortium, Rotimi, C., Abayomi, A., Abimiku, A., Adabayeri, V. M., Adebamowo, C., Adebiyi, E., Ademola, A. D., et al. (2014). Research capacity. Enabling the genomic revolution in Africa. *Science, 344*(6190), 1346–1348. https://doi.org/10.1126/science.1251546

14. Klingström, T., Mendy, M., Meunier, D., Berger, A., Reichel, J., Christoffels, A., Bendou, H., Swanepoel, C., Smit, L., Mckellar-Basset, C., & Bongcam-Rudloff, E. (2016). Supporting the development of biobanks in low and medium income countries. *2016 IST-Africa week conference 2016* May 11, 1–10. https://doi.org/10.1109/ISTAFRICA.2016.7530672

15. Abayomi, A., Gevao, S., Conton, B., Deblasio, P., & Katz, R. (2016). African civil society initiatives to drive a biobanking, biosecurity and infrastructure development agenda in the wake of the West African Ebola outbreak. *The Pan African Medical Journal, 24*(270). https://doi.org/10.11604/pamj.2016.24.270.8429

16. Hardy, B. J., Séguin, B., Goodsaid, F., & Jimenez-Sanchez-G., Singer, P.A., Daar, A.S. (2008). The next steps for genomic medicine: Challenges and opportunities for the developing world. *Nature Reviews Genetics, 9,* 23–27. https://doi.org/10.1038/nrg2444

17. Christoffels, A., & Abayomi, A. (2020). Careful governance of African biobanks. *The Lancet, 395,* 29–30. https://doi.org/10.1016/S0140-6736(19)32624-8

18. Sepulveda, J. L., & Young, D. S. (2013). The ideal laboratory information system. *Archives of Pathology & Laboratory Medicine, 137*(8), 1129–1140. https://doi.org/10.5858/arpa.2012-0362-RA

19. Kyobe, S., Musinguzi, H., Lwanga, N., Kezimbira, D., Kigozi, E., Katabazi, F.A., Wayengera, M., Joloba, M.L., Abayomi, E.A., Swanepoel, C., Abimiku, A., Croxton, T., Ozumba, P., Thankgod, A., Cristoffels, A., van Zyl, L., Mayne, E.S., Kader, M., et al. H3Africa Biorepository PI Working Group. (2017). Selecting a laboratory information management system for biorepositories in low- and middle-income countries: The H3Africa experience and lessons learned. *Biopreservation and Biobanking, 15*(2), 111–115. https://doi.org/10.1089/bio.2017.0006

20. Bendou, H., Sizani, L., Reid, T., Swanepoel, C., Ademuyiwa, T., Merino-Martinez, R., Müller, H., Abayomi, A., & Christoffels, A. (2017). Baobab laboratory information management

system: Development of an open-source laboratory information management system for biobanking. *Biopreservation and Biobanking, 15*(2), 116–120. https://doi.org/10.1089/bio.2017.0014

21. Aldosari, B., Gadi, H. A., Alanazi, A., & Househ, M. (2017). Surveying the influence of laboratory information system: An end-user perspective. *Informatics in Medicine Unlocked, 9*, 200–209. https://doi.org/10.1016/j.imu.2017.09.002

22. Argento, N. Institutional ELN/LIMS deployment: Highly customizable ELN/LIMS platform as a cornerstone of digital transformation for life sciences research institutes. *EMBO Rep. 2020;21*(3), e49862. https://doi.org/10.15252/embr.201949862

23. Myers, C., Swadley, M., & Carter, A. B. (2018). Laboratory information systems and instrument software lack basic functionality for molecular laboratories. *The Journal of Molecular Diagnostics, 20*(5), 591–599. https://doi.org/10.1016/j.jmoldx

24. Quinlan, P. R., Mistry, G., Bullbeck, H., Carter, A., & CCB Working Group 3. (2014). A data standard for sourcing fit-for-purpose biological samples in an integrated virtual network of biobanks. *Biopreservation and Biobanking, 12*(3), 184–191. https://doi.org/10.1089/bio.2013.0089

25. Eklund, N., Andrianarisoa, N. H., van Enckevort, E., Anton, G., Debucquoy, A., Müller, H., Zaharenko, L., Engels, C., Ebert, L., Neumann, M., Geeraert, J., T'Joen, V., Demski, H., Caboux, É., Proynova, R., Parodi, B., Mate, S., van Iperen, E., Merino-Martinez, R., & Silander, K. (2020). Extending the minimum information about BIobank data sharing terminology to describe samples, sample donors, and events. *Biopreservation and Biobanking, 18*(3), 155–164. https://doi.org/10.1089/bio.2019.0129

26. Hartman, V., Matzke, L., & Watson, P. H. (2019). Biospecimen complexity and the evolution of biobanks. *Biopreservation and Biobanking, 17*(3), 264–270. https://doi.org/10.1089/bio.2018.0120

27. Coppola, L., Cianflone, A., Grimaldi, A. M., Incoronato, M., Bevilacqua, P., Messina, F., Baselice, S., Soricelli, A., Mirabelli, P., & Salvatore, M. (2019). Biobanking in health care: Evolution and future directions. *Journal of Translational Medicine, 17*(1), 172. https://doi.org/10.1186/s12967-019-1922-3

The Importance of Cancer Biobanks in Low- and Middle-Income Countries

21

Io Hong Cheong and Zisis Kozlakidis

Abstract

Improving the quality of healthcare is an important goal for the global health research and wellness research system. Every patient has the right to receive timely, safe and efficient care; to be adequately informed about the care process; and to be educated about the relative risks and benefits of each treatment decision. Poor quality healthcare is wasteful and expensive and often has the greatest impact in economies and societies with the greatest need and least capital. Therefore, numerous initiatives show that the more well-characterised, high-quality samples are available through biobanks in LMICs; the faster research will advance and impact the delivery of healthcare solutions. Thus, biobanking is a key element for the success of future healthcare plans. There is thus an urgent need to improve the existing research infrastructure, including biobanks.

Keywords

Cancer biobanks · Low- and middle-income countries · World Health Organization (WHO) · International Agency for Research on Cancer (IARC)

The improvement of healthcare represents a vital objective for medical research and healthcare systems worldwide. Every patient has the right to receive timely and effective care; to be informed about the care process in a proportionate and appropriate way; and, importantly, to be informed about the relative risks and benefits of

I. H. Cheong
School of Public Health, Shanghai Jiao Tong University School of Medicine, Shanghai, China

Z. Kozlakidis (✉)
International Agency for Research on Cancer, World Health Organization, Lyon, France
e-mail: kozlakidisZ@iarc.fr

147

any actions or treatments. The provision of poor-quality healthcare can be wasteful and expensive and often has the greatest impact in economies and societies with the greatest need and the least capital.

Taking the above into consideration, in 2012 all members of the World Health Organization (WHO) approved the Global Monitoring Framework on Noncommunicable Diseases (NCDs). This framework includes a commitment to reduce premature NCD deaths (including cancer) by 25% by 2025 [1]. Furthermore, at the Seventieth World Health Assembly in May 2017, governments from around the world adopted a cancer resolution (WHA70.12): Cancer prevention and control in the context of an integrated approach [2]. This international mobilization is based on growing evidence, suggesting that most new cancer cases now occur in LMICs, increasing from 15% (1970) to 56% (2008), and expected to reach nearly 70% by 2030 [3, 4]. Given these stark figures, there is an urgent need to improve existing research infrastructure, including biobanks, ensuring that patient cancer samples and associated data are consistently of high-quality and available to researchers.

The term biobank commonly refers to a large, organised collection of well-characterised tissue samples such as surgical biopsies (freshfrozen or in paraffin sections), blood and serum samples, different cell types and DNA—all carefully collected for research purposes with their associated research and/or clinical data [5]. Within the LMICs context, most of the early and current biobanking activities, for example, in sub-Saharan Africa, are the result of studies in emerging infectious diseases, [6–8] as well as longer-term issues such as tuberculosis and malaria [9, 10]. These were "vertical" biobanks, focused on one pathology with specific pre-selection criteria for the enrolled patients and dependent on specific scientific hypotheses. As the distribution of disease types is shifting in LMICs, such "vertical" collections can become of limited scientific use—other than for the purpose they were originally created—and a wider, population-based or "horizontal approach" is required for biobanking to characterise and understand this healthcare shift within an entire population context.

For example, cancer in LMICs is characterised by increasing incidence rates of cancers with non-infectious aetiology, such as breast, prostate, lung and colorectal cancers. This shift can be explained partly by the aging of and lifestyle changes within populations, both parameters attributed at large, to economic development [11]. However, as different individuals are exposed to the above changes at different levels, and for varying lengths of time, the individual understanding of the disease based on the combination of genetic and lifestyle parameters is of paramount importance for the provision of accurate and effective medical treatment. Such precision medicine research on a population scale is based on the analyses of samples together with linked, detailed clinical data—and as signals for disease associations can be weak; thus, samples are needed in large quantities.

Therefore, there is a clear implication from the above: if larger quantities of well-characterised, high-quality samples become available through biobanks in LMICs, the faster research will be able to advance and impact upon the delivery of healthcare solutions. Thus, biobanking constitutes a critical element to the success of future healthcare plans as biobanks are now tasked with and being relied upon for the standardization of tissue collection(s) for improved science quality. Having said that, there is currently a lack of available high-quality biological resources in LMICs that could be used to better understand such observations within as well as between their populations, and link the environmental exposure to the biology of the disease at group and individual levels. Unfortunately, within resource-limited settings with large geographical areas and large populations (e.g. Ukraine, Nigeria, Egypt, Indonesia), the existence of biobanks is exceptionally rare. As such, the risk remains that if more LMICs are not represented in national and international research initiatives, they might miss opportunities that impact heavily on their communities and lose the opportunity to benefit from this research.

In one of the first surveys ever conducted among LMICs in Africa, Asia and Europe, the IARC determined that although there were some exceptions, in general biobanking in LMICs were lacking in the technical and ethical regulatory standards and infrastructure practiced in high-income settings [12]. As a result, IARC published the *Common minimum technical standards and protocols for biobanks dedicated to cancer research* so that there is a global guideline, with sections dedicated to LMICs, and additionally established in 2013 the Biobank and Cohort building Network (BCNet) [13]. BCNet is an opportunity for biobanks in LMICs to jointly address many of the challenges in biobanking and research infrastructure, the acquisition and maintenance of high-quality samples and data, and governance and regulatory frameworks to guide in the sharing and reuse of resources for research [14]. Consequently, a number of research articles on the subject were published by BCNet members [15–21].

However, beyond the apparent statements supporting the need for further development of the research capacity locally within LMICs, and the need of consistent financial support, the long-term creation and operation of biobanks within an LMIC context started to be described comprehensively relatively recently, both in operational and financial terms [22–26]. There still exists a knowledge gap for the practical implementation of long-term biobanking plans that are not necessarily related to one specific infectious disease outbreak, but of a wider nature, supporting population-based cohorts and research. Additionally, the costs often quoted for biobanks in LMICs are representative of the basal rate for fixed cost operations [22]. For example, cost models are only very few and do not include further in-kind contributions or additional variable costs, such as downstream research activities or dependencies that are typically constrained in resource-limited settings (e.g., availability of trained staff).

Therefore, a substantial challenge still exists in determining the true and representative operational costs of biobanks within an LMIC context, such that can be used as part of governmental long-term infrastructure investments and becoming eventually comparable regionally and internationally. Additionally, the international scientific literature offers very few examples on specific cost and accounting models for such organizations [27, 28]. A biospecimen user calculator tool developed by the Canadian Tissue Repository Network (CTRNet), a cancer biobank network, was published online [29, 30]. Also the National Institutes of Health, National Cancer Institute's Biorepositories and Biospecimen Research Branch, developed the

BEMT, which creates a cost profile for their biobanks' specimens, products, and services, it also established pricing and allocated costs for biospecimens [31]. Although this is a useful tool, it may not be suitable for LMICs. Thus, the development of cost models specific to cancer biobanks in LMICs is sorely needed to enable informed decision making by relevant national government and research bodies.

Future Perspectives

As a number of initiatives aim to limit the rising cancer burden in LMICs, it remains of utmost importance that the adequate application of existing knowledge takes place regarding cancer prevention, as well as the generation of new evidence [32]. Providing such evidence is a substantial opportunity for biobanks to become part of the healthcare fabric of LMICs, as opposed to study-related facilities. This paradigm shift from an "island of opportunity" to a "systemically integrated facility" is imperative to allow for long-term planning and protect the sustainability of investment.

Establishing a biobank de novo is very difficult, especially in LMIC settings, where the supporting physical infrastructure might not be amenable or even existing. Thus, setting up a biobank through collaborations with a more developed institution (s) with similar interests helps in establishing a successful biobank and lowering the risk of the overall project stalling. One such example is observed in H3Africa, where the African Society for Human Genetics partnered with the United Kingdom's Wellcome Trust and the United States' National Institutes of Health to support the establishment of biobanking activities in Africa, primarily in support of health-related genomic research [33]. Another such example of successful international cooperation is the Ukraine American cohort study sponsored by the US NCI [34].

As the concept of biobanking develops in LMICs from a state of infancy towards that of active development in international research, there need to be efforts to refine and review guidelines, regulations and laws, so that the development of biobanking and consequently of local research capacity is supported [35, 36]. Often guidelines and standard operating procedures for establishing a biobank may also not be available for the country or institution [37]. To counterbalance these needs, LMIC biobanks are increasingly more active in biobanking networks, sharing the experience and expertise with others from across the world. For example, creating national biobank networks in Indonesia and the Ukraine, as well as participating in the activities of the International Society of Biological and Environmental Repositories (ISBER) and/or BCNet. Such activities allow for expertise and information to be exchanged, strong ties to be formed as a precondition of future collaborative research [38]. These activities are expected to increase further in the future, reflecting the demand from LMIC settings.

Finally, based on the strong evidence that diagnosing many cancer types early can reduce mortality rates, primarily through the initiation of appropriate and adequate treatment in the disease's natural progression, it is likely that cancer research will move strongly into prognostic biomarkers within LMICs. Therefore, biobanks should consider identifying prognostic and diagnostic biomarkers by collecting tissue samples from individuals in at-risk population but without any clinically evident disease [39]. In the future, it is expected that the recognition of sub-clinical periods of disease development will help prevent cancer formation, regress the growth and also aid in designing new drug targets.

Conclusion

Despite the challenges and limitations mentioned previously, it is clear that there is a need for further actions across the cancer research and healthcare continuum, including for the creation of more population and/or cancer biobanks and the further strengthening of existing ones, in order to meet the demands of the increasing cancer research and incidence across LMICs. The further creation and implementation of such infrastructure will certainly require available human and financial capital, and sufficient training capacity. However, biobanks and in particular biobank networks can be positioned as a valued healthcare tool, supporting the locally relevant research and as an enabling tool for the national bioeconomy. In order to achieve this, it is crucial to engage all stakeholders, especially the clinical staff in the shared task of improving the quality of care and as a reference point for the dissemination and sharing of accrued knowledge and best practices. Tailoring the integration of biobanks to the local healthcare context may result in research outcomes that are both theoretically optimal but also practically feasible to implement.

Disclaimer Where authors are identified as personnel of the International Agency for Research on Cancer/WHO, the authors alone are responsible for the views expressed in this article and they do not necessarily represent the decisions, policy or views of the International Agency for Research on Cancer/WHO.

References

1. A Comprehensive Global Monitoring Framework, Including Indicators, and a Set of Voluntary Global Targets for the Prevention and Control of Noncommunicable Diseases. (2012). WHO. Accessed September 14, 2016, from http://www.who.int/nmh/events/2012/discussion_paper3.pdf
2. World Health Assembly, 70. (2017). *Cancer prevention and control in the context of an integrated approach.* World Health Organization. https://apps.who.int/iris/handle/10665/275676
3. World Health Organization. *Cancer: Fact sheets.* Geneva: WHO. Accessed 29 November 2013.
4. Farmer, P., Frenk, J., Knaul, F. M., Shulman, L. N., Alleyne, G., Armstrong, L., Atun, R., Blayney, D., Chen, L., Feachem, R., Gospodarowicz, M., Gralow, J., Gupta, S., Langer, A., Lob-Levyt, J., Neal, C., Mbewu, A., Mired, D., Piot, P., et al. (2010). Expansion of cancer care and control in countries of low and middle income countries: A call to action. *The Lancet, 376*(9747), 1186–1193. https://doi.org/10.1016/S0140-6736(10)61152-X

5. Mascalzoni, D. (Ed.). (2015). Biobanks: A definition. In *Ethics, law and governance of biobanking*. The International Library of Ethics, Law and Technology, 14. Springer Science and Business Media. https://doi.org/10.1007/978-94-017-9573-9

6. Abayomi, A., Gevao, S., Conton, B., Deblasio, P., & Katz, R. (2016). African civil society initiatives to drive a biobanking, biosecurity and infrastructure development agenda in the wake of the West African Ebola outbreak. *The Pan African Medical Journal, 24*(270). https://doi.org/10.11604/pamj.2016.24.270.8429

7. Franco, J. R., Simarro, P. P., Diarra, A., Ruiz-Postigo, J. A., & Jannin, J. G. (2012). The human African trypanosomiasis specimen biobank: A necessary tool to support research of new diagnostics. *PLoS Neglected Tropical Diseases, 6*(6), e1571. https://doi.org/10.1371/journal.pntd.0001571

8. Abayomi, A., Christoffels, A., Grewal, R., Karam, L. A., Rossouw, C., Staunton, C., Swanepoel, C., & van Rooyen, B. (2013). Challenges of biobanking in South Africa to facilitate indigenous research in an environment burdened with human immunodeficiency virus, tuberculosis, and emerging noncommunicable diseases. *Biopreservation and Biobanking, 11*(6), 347–354. https://doi.org/10.1089/bio.2013.0049

9. Betsou, F., Parida, S. K., & Guillerm, M. (2011). Infectious diseases biobanking as a catalyst towards personalized medicine: Mycobacterium tuberculosis paradigm. *Tuberculosis, 91*(6), 524–532. https://doi.org/10.1016/j.tube.2011.07.006

10. Sirugo, G., Van Der Loeff, M. S., Sam O, Nyan, O., Pinder, M., Hill, A. V., Kwiatkowski, D., Prentice, A., de Toma, C., Cann, H. M., & Diatta, M. (2004). A national DNA bank in the Gambia, West Africa, and genomic research in developing countries. *Nature Genetics, 36*(8), 785–786. https://doi.org/10.1038/ng0804-785

11. Kingham, T. P., & Wong, S. L. (2015). Global surgical oncology: Addressing the global surgical oncology disease burden. *Annals of Surgical Oncology, 22*(3):708–709. https://doi.org/10.1245/s10434-014-4347-5

12. Mendy, M., Caboux, E., Sylla, B., Dillner, J., Chinquee, J., & Wild, C. (2014). BCNet survey participants. Infrastructure and facilities for human biobanking in low- and middle-income countries: A situation analysis. *Pathobiology, 81*, 252–260. https://doi.org/10.1159/000362093

13. Mendy, M., Caboux, E., Lawlor, R. T., Wright, J., & Wild, C. P. (2017). *Common minimum technical standards and protocols for biobanks dedicated to cancer research*. Lyon: IARC Technical Publication No. 44.

14. Biobank and population Cohort building Network. International Agency for Research on Cancer (IARC). World Health Organization (WHO). Accessed October 20, 2020, from https://bcnet.iarc.fr/

15. Klingström, T., Mendy, M., Meunier, D., Berger, A., Reichel, J., Christoffels, A., Bendou, H., Swanepoel, C., Smit, L., Mckellar-Basset, C., & Bongcam-Rudloff, E. (2016). Supporting the development of biobanks in low and medium income countries. *2016 IST-Africa week conference 2016 May 11*, 1–10. https://doi.org/10.1109/ISTAFRICA.2016.7530672

16. Zawati, M. N., Tasse, A. M., Mendy, M., Caboux, E., Lang, M. & on Behalf of Biobank and Cohort Building Network Members. (2018). Barriers and opportunities in consent and access procedures in low-and middle-income country biobanks: Meeting notes from the BCNet training and general assembly. *Biopreservation and Biobanking, 16*(3), 171–178. https://doi.org/10.1089/bio.2017.0081

17. Stefan, D. C., & Seleiro, E. (2015). International collaboration in cancer research. In D. C. Stefan (Ed.), *Cancer research and clinical trials in developing countries* (pp. 191–199). Springer.

18. Matimba, A., Tybring, G., Chitereka, J., Zinyama-Gutsire, R., Dandara, C., Bürén, E., Dhoro, M., & Masimirembwa, C. (2016). Practical approach to biobanking in Zimbabwe: Establishment of an inclusive stakeholder framework. *Biopreservation and Biobanking, 14*(5), 440–446. https://doi.org/10.1089/bio.2015.004

19. Abd El-Aal, W. E., Abaas, N. F., El-Sharkawy, S. L., & Badawi, M. A. (2016). Biobanking: A challenge facing pathologists in Egypt. *Journal of Advances in Medicine and Medical Research, 13*(1), 1–10. https://doi.org/10.9734/BJMMR/2016/21

20. Mendy, M., Lawlor, R. T., van Kappel, A. L., Riegman, P. H., Betsou, F., Cohen, O. D., & Henderson, M. K. (2018). Biospecimens and biobanking in global health. *Clinics in Laboratory Medicine, 38*(1), 183–207. https://doi.org/10.1016/j.cll.2017.10.015

21. Afifi, N. M., Anisimov, S. V., Aguilar-Quesada, R., Kinkorova, J., Marrs, S., Nassimbwa, S., Kozlakidis, Z., & Parry-Jones, A. (2020). Biobanking spotlight on Europe, Middle East, and Africa: Presenting the collective experience of the ISBER-EMEA regional ambassadors. *Biopreservation and Biobanking, 18*(5), 471–478. https://doi.org/10.1089/bio.2020.0013

22. Fachiroh, J., Dwianingsih, E. K., Wahdi, A. E., Pramatasari, F. L., Hariyanto, S., Pastiwi, N., Yunus, J., Mendy, M., Scheerder, B., & Lazuardi, L. (2019). Development of a biobank from a legacy collection in Universitas Gadjah Mada, Indonesia: Proposed approach for centralized biobank development in low-resource institutions. *Biopreservation and Biobanking, 17*(5), 387–394. https://doi.org/10.1089/bio.2018.0125

23. Fachiroh, J. (2020). Biobank ensuring sustainability in health research. *Journal Kedokteran dan Kesehatan Indonesia, 11*(1), 3–5. https://doi.org/10.20885/JKKI.Vol11.Iss1.art2

24. Tamang, M. K., & Yadav (2020). U.N. Establishing Bio-bank in Nepal. *Journal of Nepal Health Research Council, 18*(2), 335–336. https://doi.org/10.33314/jnhrc.v18i2.2208

25. Khabour, O. F., & Abu-Siniyeh, A. (2019). Challenges that face the establishment of diabetes biobank in Jordan: A qualitative analysis of an online discussion forum. *Journal of Multidisciplinary Healthcare, 12*, 229–234. https://doi.org/10.2147/JMDH.S194161

26. Abdelhafiz, A. S., Sultan, E. A., Ziady, H. H., Ahmed, E. O., Khairy, W. A., Sayed, D. M., Zaki, R., Fouda, M. A., & Labib, R. M. (2019). What Egyptians think. Knowledge, attitude, and opinions of Egyptian patients towards biobanking issues. *BMC Medical Ethics, 20*(57). https://doi.org/10.1186/s12910-019-0394-6

27. Gonzalez-Sanchez, M. B., Lopez-Valeiras, E., Morente, M. M., & Fernandez Lago, O. (2013). Cost model for biobanks. *Biopreservation and Biobanking, 11*(5), 272–277. https://doi.org/10.1089/bio.2013.0021

28. Wu, M., Wu, D., Hu, C., & Yan, C. (2020). How to make a cost model for the birth cohort biobank in China. *Frontiers in Public Health, 8*(24). https://doi.org/10.3389/fpubh.2020.00024

29. Matzke, L., Dee, S., Bartlett, J., Damaraju, S., Graham, K., Johnston, R., Mes-Masson, A. M., Murphy, L., Shepherd, L., Schacter, B., & Watson, P. H. (2014). A practical tool for modeling biospecimen user fees. *Biopreservation and Biobanking, 12*(4), 234–239. https://doi.org/10.1089/bio.2014.0008

30. Canadian Tissue Repository Network (CTRNet). (2020). Accessed October 21, 2020, from https://www.ctrnet.ca/

31. Odeh, H., Miranda, L., Rao, A., Vaught, J., Greenman, H., McLean, J., Reed, D., Memon, S., Fombonne, B., Guan, P., & Moore, H. M. (2015). The biobank economic modeling tool (BEMT): Online financial planning to facilitate biobank sustainability. *Biopreservation and Biobanking, 16*(6), 421–429. https://doi.org/10.1089/bio.2015.0089

32. Sankaranarayanan, R., & Boffetta, P. Research on cancer prevention, detection and management in low- and medium income countries. *Annals of Oncology, 21*(10), 1935–1943. https://doi.org/10.1093/annonc/mdq049

33. Lassalle, S., Hofman, V., Ilie, M., Butori, C., Bonnetaud, C., Gaziello, M. C., Selva, E., Gavric-Tanga, V., Castillo, L., Santini, J., Chabannon, C., & Hofman, P. (2011). Setting up a prospective thyroid biobank for translational research: Practical approach of a single institution (2004–2009, Pasteur hospital, Nice, France). *Biopreservation and Biobanking, 9*(1), 9–19. https://doi.org/10.1089/bio.2010.0024

34. Devereux, L., Watson, P. H., Mes-Masson, A. M., Luna-Crespo, F., Thomas, G., Pitman, H., Speirs, V., Hall, A. G., Bollinger, N., Posada, M., & Lochmüller, H. (2019). A review of international biobanks and networks: Success factors and key benchmarks—A 10-year retrospective review. *Biopreservation and Biobanking, 17*(6), 512–519. https://doi.org/10.1089/bio.2019.29060.djc.SI

35. Nansumba, H., Ssewanyana, I., Tai, M., & Wassenaar, D. (2020). Role of a regulatory and governance framework in human biological materials and data sharing in National Biobanks: Case studies from biobank integrating platform, Taiwan and the National Biorepository, Uganda. *Wellcome Open Research, 4*(171). https://doi.org/10.12688/wellcomeopenres.15442.2

36. El-Khadry, S. W., Abdallah, A. R., Yousef, M. F., Ezzat, S., & Dorgham, L. S. (2020). Effect of educational intervention on knowledge and attitude towards research, research ethics, and biobanks among paramedical and administrative teams in the National Liver Institute, Egypt. *Egyptian Liver Journal, 10*(1). https://doi.org/10.1186/s43066-019-0010-y

37. Sajo, M. E., Teves, J. M., Corachea, A. J., Diaz, L. A., Chan, A. F., Valparaiso, A. P., Dy Echo, A. V., Macalindong, S. S., Uy, G. L., Dofitas, R. B., Habana, M. A., Gerona, R. R., Juan, C. I., Giudice, L. C., & Velarde, M. C. (2020). A pilot cancer-phenome biobanking system in a low-resource southeast Asian setting: The Philippine general hospital biobank experience. *Biopreservation and Biobanking, 18*(3), 180–188. https://doi.org/10.1089/bio.2019.0114

38. Henderson, M. K., & Kozlakidis, Z. (2018). ISBER and the biobanking and cohort network (BCNet): A strengthened partnership. *Biopreservation and Biobanking, 16*(5), 393–394. https://doi.org/10.1089/bio.2018.29043.mkh

39. Kinkorová, J. (2016). Biobanks in the era of personalized medicine: Objectives, challenges, and innovation. *EPMA Journal, 7*(4). https://doi.org/10.1186/s13167-016-0053-7

Biobanking in Paediatrics: A Look from Armenia as a Middle-Income Country

22

Sergey Sargsyan and Marine Melkumova

Abstract

Today we are witnessing significant advances in various fields of medicine. In particular, many of the recent advances in biomedical research are largely related to advances in sequencing the human genome, identifying the role of specific genes in the development of certain diseases and genomics studies. Some of these studies have been put into practice and have led to dramatic improvements in the treatment of disease groups. In addition, the scientific community has taken further important steps towards more intensive and productive research in the field of human genetics over the past decade, including the establishment of repositories for human DNA and other materials. This chapter addresses questions such as "Why should we think about biobanking, what will be different for the children we care for? Should we expect progress in care? For those of us who are representatives of the paediatric system in a middle-income country, what are the settings and context of paediatric biobanking? What will be unique about biobanking in poorer countries?"

Keywords

Paediatrics · Low- and middle-income countries · Disease groups · Human genetics

Nowadays we are witnessing significant progress in different fields of medicine. In particular, many recent advances in biomedical science are largely associated with the progress of sequencing of the human genome, the establishment of the role of specific genes, as well as in the development of particular diseases, as well as studies

S. Sargsyan (✉) · M. Melkumova
Arabkir Medical Centre, Yerevan, Armenia

in the field of genomics. Some of these studies have been put into practice and led to dramatic improvements in the management of disease groups. In addition, over the last decade, the scientific community has taken other major steps towards more intensive and productive research in the field of human genetics, predominantly through the establishment of repositories for human DNA and other materials.

Why should we think about biobanking, what will be different for the children we care for? Should we expect any progress in care? For those of us who are representatives of the paediatric system in a middle-income country, what are the settings and the context of paediatric biobanking?

To better understand the answers, two aspects of the question should be analysed: The first aspect is "paediatric" vs. "adult"; the second one is related to resource availability, e.g., "process in high-income countries" vs. "process in low- and middle-income countries."

"Children Versus Adults"

So far, most studies generated by biobanks worldwide have focused on non-infectious diseases in adults. This is quite understandable, as these diseases are responsible for the majority of morbidity and mortality and the overall burden of disease worldwide.

The situation with children and adolescents generally and in regard of biobanking particularly is different as children truly cannot be considered as "little adults" [1]. The main causes of mortality and morbidity are related to genetic and congenital factors and infectious diseases. Consequently, many studies related to the genetics and phenotype of children in the first years of life are primarily related to genetic causes of disease, environmental factors and resistance to infectious agents.

Since many genetic diseases manifest at a young age, sick children and their families, paediatricians and the entire child health care system worldwide are among the beneficiaries of advances in biomedical, especially genetic, research. Genetic analysis of specific mutations, performed in accordance with comprehensive clinical assessment and biomarker testing, helps clinicians provide a much more efficient diagnosis, treatment and prognosis for patients with, e.g. cystic fibrosis, lysosomal storage diseases, some neurodevelopmental disorders and numerous other conditions. Therefore, it is clear that studies promoted by paediatric biobanks should also include the specific "paediatric" topics.

One possible area of research is, for instance, the investigation of rare diseases, especially genetic syndromes. These studies are often difficult as the limited number of observations in a certain research centre may not be adequate in order to make relevant conclusions on their etiology and pathogenesis and develop evidence-based treatment methods. The preservation of biomaterials (including post-mortem), further sharing with other centres, and accumulation will lead to larger populations to draw some consistent conclusions, as well as preserving these materials for the future when some other methods may emerge and bring benefits to research and practical medicine.

The prevalence and structure of infectious diseases in children are somewhat different from adults. The response mechanisms of the child's body to infectious agents are also dissimilar, leading to different outcomes for certain infections; the current coronavirus pandemic is a case in point. Biobanking may broaden the spectrum of studies, improve the understanding of these variations and facilitate further studies and eventual preventive measures. On the other hand, many of the above-mentioned noncommunicable diseases in adults have their origin in childhood. Genetic predisposition, environmental influences (including antenatal period), genotype-phenotype correlations in cardiovascular, oncological and other diseases are the research fields in which a biobanking-based longitudinal approach can enrich current studies and make an important contribution.

Apart from the specificity of paediatric topics, another obvious problem with paediatric biobanking is the consent issue [2]. Appropriate regulatory frameworks are needed to regulate the mechanisms for obtaining parental consent, the need to continue or renew consent in the event of changes in the status of the child's legal ties, and the attainment of majority. A specific problem is the definition of the moment of legal responsibility for adolescents. In many countries, adolescents have the right to access medical facilities with no parental consent before reaching legal age, at least for some types of care. For example, according to the National Reproductive Law, Armenian adolescents have a legal right; therefore, from a medical and legal point of view, sample collection procedures should be thoroughly assessed and related to existing regulations.

Therefore, "paediatric" centres should operate with a long-term perspective, and multidisciplinary networks of institutions and other centres need to be established. This in turn raises questions about costs, relatively complicated operating manuals adapted to specific research needs and mechanisms of collaboration. However, the benefits of interventions may significantly outweigh the potential costs and open new horizons for studies.

"High-Income Countries Versus Low- and Middle-Income Countries"

Most of the initial studies and further translation of research into practice took place in high-income countries that are able to devote sufficient resources to science and health spending. For example, genetic studies have been invented and are currently widely accessible in most of Europe and North America. Later, these relatively expensive and sophisticated techniques will be gradually introduced into low- and middle-income countries, regardless of whether there is a partial or significant shortage of resources and personnel. Presumably, even in low-resource settings, many of these technologies bring financial benefits in addition to clinical benefits by reducing overall expenditures through optimization of patient management. However, it should be noted that the overall and long-term cost-effectiveness for the introduction of many of the modern technologies into daily practice in developing countries is not thoroughly studied yet.

What will be unique when biobanking will be used in poorer countries? A lack of resources may push the initiative groups to seek the support from different funds and technical support [3]. The research capacities in most countries may require significant technical support from existing advanced biobanking centres. As legal regulations in many of those countries in general are weaker, it may require some additional efforts and consultancy for establishing a good legal basis for the process and the covering of all ethic aspects. The national believes and the cultural specific issues should be considered as well [4]. The experiences of previous programs, especially those related with carriers of genetic information, must be taken into account.

On the other hand, what will be the positive outcome of establishing paediatric biobanks in the developing world? Benefits include reducing inequities for children, covering previously understudied populations and enlarging the volume and spectrum of biological information available to researchers for analysis. Finally, yet importantly, it improves the research capacity in low- and middle-income countries, which can have secondary positive outcomes in other fields as well [3–5].

To understand the positive results of introducing genetic studies in low-to-middle income country better, the case of Armenia can be considered. The Republic of Armenia, which gained independence in 1991, inherited a relatively well-developed public health system and some research capacity from the Soviet era. However, the health expenditures in Armenia were, and in some extent still are, very low and even among the lowest worldwide, due to the severe economic crisis after the collapse of the Soviet Union and the disruption of traditional economic ties. Like all health sectors, Armenia's child health system is underfunded but reasonably efficient, as indicated by the relatively low child mortality rate, which was 8.6 per 1000 children in 2019 [6].

One of the key priority research topics for Armenia is Familial Mediterranean Fever (FMF), a genetic disease that belongs to the group of periodic fevers and auto inflammatory diseases and affects mainly Armenians, non-Ashkenazi Jews, Arabs, Turks and only rarely other nations. The key signs of this disease are recurrent attacks of fever, abdominal and thoracic pain and arthritis; the most dangerous complication is amyloid nephropathy, which leads to chronic renal failure. Until the 1990s, in Armenia as in other countries, the diagnosis of FMF was made only based on clinical criteria, which was not so difficult in the case of typical manifestations. The gene responsible for developing FMF (MEFV) was discovered in 1997; the testing for it has been introduced in Armenia in the early 2000s, thanks to the efforts of the National Centre of Medical Genetics. Genetic testing, especially in abortive and atypical manifestations with the phenotype of the disease, genetic testing for the risk of nephropathy, carried out in line with establishing effective clinical services, providing free drugs and regular follow-up for paediatric patients, made a significant contribution to improving the management of patients and preventing complications [7]. The assessment of effectiveness of spent financial resources (including state financing of genetic testing) demonstrated the high cost-effectiveness of these measures.

This example shows that in a country with scarce resources, the introduction of modern and relatively expensive technologies into medical practice can be a resource-saving intervention. The same can be expected from paediatric biobanks. If we understand the essence of this process, the establishment of biobanks and the rational organization of the recruitment of materials will provide new opportunities for studies on the infant body, which in turn will increase the opportunities for translating the results of these studies into daily practice.

References

1. Catchpoole, D. R., Carpentieri, D., Vercauteren, S., Wadhwa, L., Schleif, W., Zhou, L., Zhou, J., Labib, R. M., Smits, E., & Conradie, E. H. (2020). Pediatric biobanking: Kids are not just little adults. *Biopreservation and Biobanking, 18*(4), 258–265. https://doi.org/10.1089/bio.2020.29071.djc
2. Moradian, M. M., Sarkisian, T., Ajrapetyan, H., & Avanesian, N. (2010). Genotype–phenotype studies in a large cohort of Armenian patients with familial Mediterranean fever suggest clinical disease with heterozygous MEFV mutations. *Journal of Human Genetics, 55*, 389–393. https://doi.org/10.1038/jhg.2010.52
3. Ndebele, P., & Musesengwa, R. (2008). Will developing countries benefit from their participation in genetics research? *Malawi Medical Journal, 20*(2), 67–69. https://doi.org/10.4314/mmj.v20i2.10960
4. Sargsyan, S., Movsesyan, Y., Melkumova, M., & Babloyan, A. (2016). Child and adolescent health in Armenia: Experiences and learned lessons. *The Journal of Pediatrics, 177*, 21–34. https://doi.org/10.1016/j.jpeds.2016.04.038
5. Séguin, B., Hardy, B. J., Singer, P. A., & Daar, A. S. (2008). Genomic medicine and developing countries: Creating a room of their own. *Nature Reviews Genetics, 9*, 487–493. https://doi.org/10.1038/nrg2379
6. Thong, M. K., See-Toh, Y., Hassan, J., & Ali, J. (2018). Medical genetics in developing countries in the Asia-Pacific region: Challenges and opportunities. *Genetics in Medicine, 20*, 1114–1121. https://doi.org/10.1038/s41436-018-0135-0
7. Hens, K., Van El, C. E., Borry, P., Cambon-Thomsen, A., Cornel, M. C., Forzano, F., Lucassen, A., Patch, C., Tranebjaerg, L., Vermeulen, E., Salvaterra, E., Tibben, A., Dierickx, K., & PPPC of the European Society of Human Genetics. (2013). Developing a policy for paediatric biobanks: Principles for good practice. *European Journal of Human Genetics, 21*(1), 2–7. https://doi.org/10.1038/ejhg.2012.99

Biobanking of the Central Nervous System **23**

Tanja Macheiner, Christine Mitchell, Konstantin Yenkoyan, Armen Muradyan, and Karine Sargsyan

Abstract

Brain tissue biobanks represent specific collections and ensure investigation and research into neurological and psychiatric diseases. Depending on the research question, scientists can use post-mortem biospecimens or, more rarely, samples from living donors. Generally, there are post-mortem samples, which are retrieved during autopsy. However, any extent of post-mortem delay until or after autopsy affects tissue analyses such as immunohistochemical labelling. Currently, the European Brain Bank Network for neurobiological studies in neurological and psychiatric diseases comprises nine partner institutions.

Keywords

Brain tissue biobank · Brain bank network · Neurological and psychiatric diseases · SOP

Brain tissue biobanks represent specific collections and ensure investigation and research into neurological and psychiatric diseases. Depending on the research question (hypothesis or specific scientific topic), scientists can use post-mortem

T. Macheiner · C. Mitchell
International Biobanking and Education, Medical University of Graz, Graz, Austria

K. Yenkoyan · A. Muradyan
Yerevan State Medical University, Yerevan, Armenia

K. Sargsyan (✉)
International Biobanking and Education, Medical University of Graz, Graz, Austria

Department of Medical Genetics, Yerevan State Medical University, Yerevan, Armenia

Ministry of Health of the Republic of Armenia, Yerevan, Armenia
e-mail: karine.sargsyan@medunigraz.at

161

biospecimens or, more rarely, samples from living donors (e.g. tumour surgery leftover material). Furthermore, the sample type plays an essential and fundamental role in the analysis; therefore researchers have to choose between formalin-fixed, paraffin-embedded tissue (FFPE), paxgen-fixed, paraffin-embedded tissue (PFPE) and fresh frozen tissue samples, amongst others. FFPE tissues often form important resources of clinical biobanks for retrospective studies. As these types of collections are very expensive and intended for long-term as well as multiple use, a business plan is essential. Several standards are needed to avoid pitfalls in this biomedical research field:

1. Standardized informed consent procedure according to institutional, local, regional, national as well as international legal and ethical procedures and standards for working with human biological material
2. Rapid autopsy system to minimize autolytic artefacts
3. Compatible SOPs for management, handling and storage
4. Consistent diagnostic criteria (e.g. free text vs. ICD-10 Code)
5. Safety precautions [1]

Generally, there are post-mortem samples, which are retrieved during autopsy (so mostly no standardization on time is possible). However, any degree of post-mortem delay until or after autopsy has an impact on tissue analyses such as immunohistochemical labelling. The first histological changes can be detected as early as 10 min after death. This is followed by morphological changes in the cell organelles, such as the swelling of the mitochondria [2]. It shows how important a rapid autopsy system is for the quality of biospecimens. Therefore, post-mortem interval, RNA integrity and pH are used as biomarkers for the quality of tissue [3].

Currently, the European Brain Bank Network for neurobiological studies in neurological and psychiatric diseases comprises nine partner institutions (https://cordis.europa.eu/project/id/CIPD940236/de).

Brain banking is a separate branch of biobanking. The term "brain banking" is usually used to describe the biobanking of central nervous system (CNS) tissues (e.g. entirely retrieved brains, brain tissue pieces gained by biopsies and spinal cord samples) and associated fluids, e.g., cerebrospinal fluid (CSF) and blood. Liquor and CNS biopsy tissue samples can be gained during a surgery, whereas the whole brain must be taken from dead patients and healthy deceased people [4, 5]. Post-mortem brain donation may be required for neurological testing, but it raises some questions that set it apart from the biobank system of other tissue samples to the collection and handling of human brain tissue. The whole brain forms the core stone for research that can explore complex brain diseases such as neurodegenerative diseases. On other hand, the animal models are useful for these types of diseases, but human tissue biomaterials as well as cerebrospinal fluid from patients with brain disease cannot be replaced. Since CSF banks are not very different from other biological fluid banks, here we focus on tissue preservation, especially of the whole brain.

Following the global expansion of aging societies, healthcare expenses will geometrically continue to rise over the next few decades [6]. According to a 2004 study by the *European Brain Council*, 127 million Europeans experience brain diseases and spend about USD 400 billion on the healthcare system of an aging society [7]. Research into the brain and CSF fluid can bring better understanding, treatment, diagnostic and prevention in all diseases of the CNS [7]. Current research activities worldwide are focused on comparatively widespread disease-specific health conditions. Diseases such as Parkinson's, HIV or mental illness are very rare in the field of biobanking [8–10]. Some brain banks operate on a large scale; for example, the brain bank in the Netherlands has received tissue from over 3700 brain donors starting from 1985, and the New York brain bank has brains from over 5000 donors [11, 12].

Because of the understandable shortage of brain donors and scientific working tendency regarding networking, in 2001, the *European Commission* presented *Brain Net Europe* [13, 14]. A consortium of international brain banks has also been founded, operating around the globe. As brain banking is a highly comprehensive field, scientific networking is required [14]. Effective networking requires standardized procedures, diagnostic methods, and information collection so that biomaterial and corresponding data can be compared and pooled in bigger cohorts and projects. That is why the uniqueness of brain banking in the context of biobanking is often highlighted, showing the differences between donations of brain tissue and other tissues/organs [15, 16]. Although several recommendations have been made for brainbank management, there are no standard protocols or guidelines for brain tissue storing worldwide [17–19]. The Brain Net Europe consortium has also published a code of conduct for ethical brain banking [20, 21].

In debates about biobanking in general, the special issues associated with post-mortem human brain tissue, and in particular the collection, processing and preservation of the whole brain have been addressed very little due to the sensitive nature of the topic. There are some very specific features that make brain banking an independent form of biobanking which combines the viewpoints of biomedical ethics, neuropathology and health system informatics. Whilst politics, practicality and data governance are often considered separately as specific aspects of brain banking, we think one should encompass the entire process. A comprehensive understanding of the challenges associated with brain banking is important to not only improve the effectiveness of the practice but correspondingly being able to counter the present-day deterioration in post-mortem brain donation frequencies.

Brain donor recruitment is now increasingly organized through donation programs, wherein those donors (mostly diseased persons with a neurodegenerative or a mental illness, but also healthy controls) can sign up to brain banks prior to their deaths. Ethically, the benefits of post-mortem donation programs include initial interactions amongst brain banks (e.g. a brain bank employee, particularly a neurologist, but also researchers) and the general population, patients with the named diagnoses, healthy controls and patient organizations. From a research point of view, donation programs are valuable as they allow the continuous collection of health information and human biological materials (such as cerebrospinal fluid, blood, etc.) throughout the donor's life. In addition, they allow autopsy, after which the data obtained during the illness can be compared with the results of the post-mortem examination.

When recruiting donors, it is necessary to document the initial diagnosis, especially if an infectious disease is suspected. Likewise, where potential donors are available, physicians should contact specialists and appropriate monitoring units for a preliminary diagnosis. There are also IT challenges in brain banking and donor recruitment. The donation programs should document the donor's medical records, contact details (and close family members) and consent forms provided by donors, for example, to conduct genetic testing or to inform about upcoming tests. Brain donations can be made long after consent is given, so every brain bank must keep this donor information up to date. The research subject administration system is an IT tool that is used for processing information that relates to the communication and consent of donors. Most donors are affected by neurodegenerative, mental or oncologic diseases, so the banked samples are indispensable for understanding the causes of mental illnesses. Very few people realize that healthy donors are also needed in brain banking as they serve as a control group. As a result, brain banks frequently reach out to partners or families of donors who show higher interest in contributing their brains to research than unaffected population, as they have a direct understanding of the importance of research in the field.

Another challenge in brain banking is the reduction of autopsies worldwide [15]. Quotas are reduced because current legal norms make it difficult to perform an autopsy, so doctors no longer want to trouble relatives with autopsies [22]. Thus, brain contribution to research advances moral issues that do not arise in other biobanking activities. Although "brain apparatus" is supported by myths, studies show that many people do not see a fundamental difference between brain donation and any other organ donation [23–25]. Nevertheless, studies show that the human population is, if directly affected, sensitive to the matter. For example, one study suggested that most respondents were less concerned when brain tissue was removed than donating the entire brain [26]. For mentally ill people who are eligible donors, it can be discomforting to join a brain bank that specializes in such conditions; even healthy donors may be uncomfortable with the idea. Accordingly, brain samples are extra challenging to obtain than any other human biological material [27].

Brain banking also faces its challenges when it comes to getting an agreement. Isaac's report in 2003 showed that in the UK several thousand brains had been collected without the family's consent. This led to the revision of the *Human Tissue Act (HTA)*, which required the consent of relatives to open or destroy organs and tissues [14, 28]. The second question concerns the acceptance of an individual whose perception is severely diminished by a mental or neurodegenerative disease (e.g. schizophrenia, Alzheimer's disease) [28]. Whereas scientific ethics offer specific support for the same disease, the patient's legal representative may consent [29, 30]. Proper planning in the case of brain donation is directly affected by the legal consent of the representative. As many neurodegenerative diseases deteriorate over time, the informed consent can be gained whilst the disease is still inactive. In addition to severe depression or psychosis, the mentally ill may also be educated about brain banking and, if having enough time, may also consent themselves to the donation [15, 31].

In general, biobanking requires more careful authorization, which is often impossible in practice. If the research project exceeds the initial approval, and if consent is not possible with the satisfied efforts, the ethics committee will evaluate by default [32]. Consent of the donor is at least theoretically possible, but post-mortem collection naturally precludes this option. Brain banking can fully ensure the initial consent of the donor and his/her immediate family so that they are not faced with any decisions in an already stressful situation. This can help ensure (often for family members if the diagnosis is hereditary) that the tissue samples and other specimens will be used in modern research that employs the latest analytical methods such as whole genome sequencing, targeted sequencing, metabolomics and other effective methods. Most importantly, even with broad consent, an independent ethics committee must review each research project.

From a research perspective, the post-mortem period should be as short as possible (less than 24 h, if possible) to ensure tissue quality. In either case, the post-mortem period must be documented to compare samples. The short time required for an autopsy (if the body has to be transported long distances to the neurology department) can lead to research and organizational problems. In particular, in the case of sudden death, the brain bank relies on the willingness of family members to immediately alarm the brain bank to deliver the obligatory documentation. Because this is stressful, some brain banks offer detailed advice on how to issue brain donation cards to potential donors and offer psychological support to the family.

If the deceased is not registered and his/her wishes have not been expressed in lifetime, it may be helpful to contact family members in advance, as autopsies require family consent in many countries. Some brain banks contact family members by telephone after death which can be a significant burden [14]. There is a very narrow and difficult margin between sensitivity and testing quality requirement [25]. Family members can say that the post-mortem examination of the deceased will not affect the changes. In many cases, once specialists interpret the family's problems seriously and offer appropriate counselling (comprising medical and scientific groups), donations can result in even greater support for the family [25, 26, 33].

From an ethical position, family members of donors have the same role in the context of donating organs for transplantation as they do in the perspective of brain contribution to biomedical research. Regardless of whether the law requires the following permission from family members, when the patient during life time has decided to make a donation, in real life, the banned family members have the decision power. Therefore, according to the Brain Net Europe model form, no autopsy is performed unless the family decisively rejects the donor's consent [34].

In fact, the position of the family is particularly prominent in the situation with brain donation. For example, in Germany, a family member, as the deceased legal representative, is anticipated to agree on biobanking contribution. According to the patient's wish, brain donations do not happen, if the representative determinedly declines. The proper balance amongst the donor's choice and the influence it has on her/his immediate family are currently being discussed, for instance, the issue if the deceased should be allowed to vote on the family order [28]. Since donors have the right to respect their own free will, scientific investigations on brain contributing to the overall good in well-being.

When the family agrees, the brain tissue is released for research. Setting a sample range and maintenance depends on the initial diagnosis and also the objective of the study. The experimental diagnosis determines if additional tissues such as the spine (more difficult to remove than the brain) are required. However, this may exceed the estimated shelf life/storage area. Consequently, project standards for specific diseases are established in prior.

Standard research data describing all stages of brain material (decomposition, method of preservation, storage) will then be collected, including the unique number of each sample, the exact storage location, and any available quality information. Another important explanatory function is to map the small parts of cells that are removed to the original part of the brain and then map all with codes [35]. Brain Net Europe has developed a special workflow and coding system for the labelling of tissue [36]. Unfortunately, the corresponding SOPs are not available for all diseases [17]. In addition, it is stated that it is very difficult to determine the minimum amount of data collected by the brain bank, especially in case of mental illnesses. However, data sharing is the most important aspect in encouraging scientists to make their data public.

It is necessary to store samples properly in order to keep the donor brain qualitatively neutral for long time storage. For this purpose, brain tissue must be widely described with high-level biological and histopathological analyses. Therefore, parts of the brain should be stored between $-150\ °C$ and $-160\ °C$ (gas phase of liquid nitrogen), whilst other parts (usually 10% for laboratory analysis, corresponding to 30–40% FFPE) should be fixed immediately. Although the Neuronal Tissue Consortium Protocol calls for coronary artery bypass coagulation, the Edinburgh Brain Bank has banned the selection of parts of HIV-infected cases, preferring more refrigeration because large-cell patches later require extensive maintenance. However, Columbia University has developed a modern protocol for the disintegration of new brain sampling which results in 1150 individual samples.

The pH value is the unit that determines the quality of stable cell samples. In direct relation to the quality of RNA, which can be measured by determining the total RNA output per unit, it is the rate of decrease of ribosomal RNA [17, 37]. The time of the patient's death should be considered when comparing model cases.

In addition to clinical interventions, researchers should ensure that the study conducted is unique and brings innovation to the field of neuroscience. To ensure that relevant specimens are searched in each affected area, Brain Net Europe provides an internal standard that includes screening, age, sex, cause of death, strong clinical and pharmacological history, agonist status, post-mortem history, and neurological diagnosis, as well as a list of relevant specimens linked to brain areas. However, the standard has two problems. Firstly, it was not clear whether this minimum data set was available to all brains from all European brain banks. The second problem is the lack of a standardized list that allows a researcher or biobanker to efficiently search for a simple database. Although these problems affect not only brain banks, it can also cause certain problems because not all brain donors are able to respond to the history or course of the disease.

Based on an analysis of brain bank profiles as opposed to a general biobank, we identified at least three issues that require further discussion. First, the brain banks and the specialists must have long-term training. Depending on the disease, the collection of donor brains can take many decades. Therefore, in the long run, diagnosis, collection of clinical data, follow-up examinations and informed consent should be continually considered. Successful brain bank research requires cross-sectional support from neurosurgeons, neurologists, therapists, ethicists, IT professionals, etc., especially at the recruitment stage. Secondly, the acquisition, processing and archiving of brain material is more time-consuming than standard biobanking, as each brain area has its specific description, which should be taken in account. Sustainable financing and cost recovery of brain banks are key issues. If public funds in the long run are not enough to cover the costs of brain banking, future brain banks will have to find models to cover their costs. Thirdly, biomarker research in neurosciences faces a lack of good quality brain material, especially well characterized tissue. A network of brain banks with a fair and efficient distribution system is very important. The acquisition of whole brain material ultimately depends on the death of the donor; therefore, special requirements must be met for its collection. In this sense, the role of the other family members deserves more attention. Some researchers argue that studies of neurodegenerative psychiatric illnesses are in the public interest not only of the family's right to revoke a donor's consent after the donor's death but also "consent in favor of the donor in the event of death". However, one must believe in the conditions of personal research and contribution to this place. Although this belief may not be an honour, it is better to talk about improving the general understanding of brain donation than to legally deny family problems.

References

1. Ravid, R., & Ikemoto, K. (2012). Pitfalls and practicalities in collecting and banking human brain tissues for research on psychiatric and neulogical disorders. *Fukushima Journal of Medical Science, 58*(1), 82–87. https://doi.org/10.5387/fms.58.82
2. Scudamore, C. L., Hodgson, H. K., Patterson, L., Macdonald, A., Brown, F., & Smitch, K. C. (2010). The effect of post-mortem delay on immunohistochemical labelling – A short review. *Comparative Clinical Pathology, 20*, 95–101. https://doi.org/10.1007/s00580-010-1149-4
3. Palmer-Aronsten, B., Sheedy, D., McCrossin, T., & Kril, J. (2016). An international survey of brain banking operation and characterization practices. *Biopreservation and Biobanking, 14*(6). https://doi.org/10.1089/bio.2016.0003
4. Dedova, I., Harding, A., Sheedy, D., Garrick, T., Sundqvist, N., Hunt, C., Gillies, J., & Harper, C. G. (2009). The importance of brain banks for molecular neuropathological research: The New South Wales tissue resource Centre experience. *International Journal of Molecular Sciences, 10*(1), 366–384. https://doi.org/10.3390/ijms10010366

5. Harmon, S. H., & Mcmahon, A. (2014). Banking (on) the brain: From consent to authorisation and the transformative potential of solidarity. *Medical Law Review, 22*(4), 572–605. https://doi.org/10.1093/medlaw/fwu011

6. Harper, S. (2014). Economic and social implications of aging societies. *Science (New York, N. Y.), 346*(6209), 587–591. https://doi.org/10.1126/science.1254405

7. Di Luca, M., Baker, M., Corradetti, R., Kettenmann, H., Mendlewicz, J., Olesen, J., Ragan, I., & Westphal, M. (2011). Consensus document on European brain research. *The European Journal of Neuroscience, 33*(5), 768–818. https://doi.org/10.1111/j.1460-9568.2010.07596.x

8. de Oliveira, K. C., Nery, F. G., Ferreti, R. E., Lima, M. C., Cappi, C., Machado-Lima, A., Polichiso, L., Carreira, L. L., Ávila, C., Alho, A. T., Brentani, H. P., Miguel, E. C., Heinsen, H., Jacob-Filho, W., Pasqualucci, C. A., Lafer, B., & Grinberg, L. T. (2012). Brazilian psychiatric brain bank: A new contribution tool to network studies. *Cell and Tissue Banking, 13*(2), 315–326. https://doi.org/10.1007/s10561-011-9258-0

9. Stone, K. (2011). Researchers take on a preventable dementia: Brain bank is giving researchers new understanding of chronic traumatic encephalopathy. *Annals of Neurology, 70*(2), A11–A14. https://doi.org/10.1002/ana.22540

10. Morgello, S., Estanislao, L., Simpson, D., Geraci, A., DiRocco, A., Gerits, P., Ryan, E., Yakoushina, T., Khan, S., Mahboob, R., Naseer, M., Dorfman, D., Sharp, V., & Manhattan HIV Brain Bank. (2004). HIV-associated distal sensory polyneuropathy in the era of highly active antiretroviral therapy: The Manhattan HIV brain Bank. *Archives of Neurology, 61*(4), 546–551. https://doi.org/10.1001/archneur.61.4.546

11. Netherlands Brain Bank. Netherlands Brain Bank Progress Report 2011–2012. (2013). Accessed July 05, 2021, from http://www.brainbank.nl/media/uploads/file/NBB-PR1112.pdf

12. Columbia Medicine. Neighbor J. (2014). *New York brain bank: Deposits and withdrawals that make a difference.* Accessed July 05, 2021, from http://www.columbiamedicinemagazine.org/features/fall-2014/new-york-brain-bank

13. Bell, J. E., Alafuzoff, I., Al-Sarraj, S., Arzberger, T., Bogdanovic, N., Budka, H., Dexter, D. T., Falkai, P., Ferrer, I., Gelpi, E., Gentleman, S. M., Giaccone, G., Huitinga, I., Ironside, J. W., Klioueva, N., Kovacs, G. G., Meyronet, D., Palkovits, M., Parchi, P., ... Kretzschmar, H. (2008). Management of a twenty-first century brain bank: Experience in the BrainNet Europe consortium. *Acta Neuropathologica, 115*(5), 497–507. https://doi.org/10.1007/s00401-008-0360-8

14. Schmitt, A., Bauer, M., Heinsen, H., Feiden, W., Consortium of Brainnet Europe II, Falkai, P., Alafuzoff, I., Arzberger, T., Al-Sarraj, S., Bell, J. E., Bogdanovic, N., Brück, W., Budka, H., Ferrer, I., Giaccone, G., Kovacs, G. G., Meyronet, D., Palkovits, M., Parchi, P., et al. (2007). How a neuropsychiatric brain bank should be run: A consensus paper of Brainnet Europe II. *Journal of Neural Transmission (Vienna, Austria: 1996), 114*(5), 527–537. https://doi.org/10.1007/s00702-006-0601-8

15. Kretzschmar, H. (2009). Brain banking: Opportunities, challenges and meaning for the future. *Nature Reviews. Neuroscience, 10*(1), 70–78. https://doi.org/10.1038/nrn2535

16. Ravid, R. (2008). Standard operating procedures, ethical and legal regulations in BTB (brain/tissue/bio) banking: What is still missing? *Cell and Tissue Banking, 9*(2), 121–137. https://doi.org/10.1007/s10561-007-9055-y

17. Ravid, R., & Park, Y. M. (2014). Brain banking in the twenty-first century: Creative solutions and ongoing challenges. *BSAM, 2014*(2), 17–27. https://doi.org/10.2147/BSAM.S46571

18. Model Brain Bank Regulations. (2009). Accessed July 05, 2021, from https://www.neuropathologie.med.uni-muenchen.de/download/infoflyer_neurobiobank_v2

19. McMahon, A., & Harmon, S. H. (2012). Banking (on) the brain: A report on the legal and regulatory concerns. *SCRIPT-ed., 9*(3), 376–383. https://doi.org/10.2966/scrip.090312.376

20. Supplier Code of Conduct. Accessed July 05, 2021, from https://www.bain.com/about/further-global-responsibility/our-sustainability/supplier-code-of-conduct

21. Klioueva, N. M., Rademaker, M. C., Dexter, D. T., Al-Sarraj, S., Seilhean, D., Streichenberger, N., Schmitz, P., Bell, J. E., Ironside, J. W., Arzberger, T., & Huitinga, I. (2015). BrainNet

Europe's code of conduct for brain banking. *Journal of Neural Transmission (Vienna, Austria: 1996), 122*(7), 937–940. https://doi.org/10.1007/s00702-014-1353-5

22. O'Grady, G. (2003). Death of the teaching autopsy. *BMJ (Clinical Research ed.), 327*(7418), 802–803. https://doi.org/10.1136/bmj.327.7418.802

23. Eatough, V., Shaw, K., & Lees, A. (2012). Banking on brains: Insights of brain donor relatives and friends from an experiential perspective. *Psychology & Health, 27*(11), 1271–1290. https://doi.org/10.1080/08870446.2012.669480

24. Olesen, J., Baker, M. G., Freund, T., di Luca, M., Mendlewicz, J., Ragan, I., & Westphal, M. (2006). Consensus document on European brain research. *Journal of Neurology, Neurosurgery, and Psychiatry, 77*(Suppl 1), i1–i49.

25. Garrick, T., Sundqvist, N., Dobbins, T., Azizi, L., & Harper, C. (2009). Factors that influence decisions by families to donate brain tissue for medical research. *Cell and Tissue Banking, 10*(4), 309–315. https://doi.org/10.1007/s10561-009-9136-1

26. Millar, T., Walker, R., Arango, J. C., Ironside, J. W., Harrison, D. J., MacIntyre, D. J., Blackwood, D., Smith, C., & Bell, J. E. (2007). Tissue and organ donation for research in forensic pathology: The MRC sudden death brain and tissue bank. *The Journal of Pathology, 213*(4), 369–375. https://doi.org/10.1002/path.2247

27. Amarasinghe, M., Tan, H., Larkin, S., Ruggeri, B., Lobo, S., Brittain, P., Broadbent, M., Baggaley, M., & Schumann, G. (2013). Banking the brain. Addressing the ethical challenges of a mental-health biobank. *EMBO Reports, 14*(5), 400–404. https://doi.org/10.1038/embor. 2013.46

28. Boyes, M., & Ward, P. (2003). Brain donation for schizophrenia research: Gift, consent, and meaning. *Journal of Medical Ethics, 29*(3), 165–168. https://doi.org/10.1136/jme.29.3.165

29. Council of Europe. (2017). *Convention for the Protection of Human Rights and Dignity of the Human Being with regard to the Application of Biology and Medicine: Convention on Human Rights and Biomedicine*. Accessed July 05, 2021, from http://conventions.coe.int/Treaty/en/Treaties/html/164.htm

30. World Medical Association. (2013). *WMA Declaration of Helsinki – Ethical Principles for Medical Research Involving Human Subjects*. Accessed 05 Jul 2021, from https://www.wma.net/wp-content/uploads/2016/11/DoH-Oct2013-JAMA.pdf

31. Hulette, C. M. (2003). Brain banking in the United States. *Journal of Neuropathology and Experimental Neurology, 62*(7), 715–722. https://doi.org/10.1093/jnen/62.7.715

32. Council of Europe. (2006). *Recommendation Rec 4 of the Committee of Ministers to member states on research on biological materials of human origin*. Accessed July 05, 2021, from https://wcd.coe.int/ViewDoc.jsp?id=977859

33. Azizi, L., Garrick, T. M., & Harper, C. G. (2006). Brain donation for research: Strong support in Australia. *Journal of Clinical Neuroscience: Official Journal of the Neurosurgical Society of Australasia, 13*(4), 449–452. https://doi.org/10.1016/j.jocn.2005.06.008

34. BrainNet Europe. (2010). *Model information leaflet concerning registration for and brain donation to the brain bank*. Accessed July 05, 2021, from https://www.neuropathologie.med.uni-muenchen.de/download/nbm_informationsblatt_kurz_1.pdf

35. Nussbeck, S. Y., Skrowny, D., O'Donoghue, S., Schulze, T. G., & Helbing, K. (2014). How to design biospecimen identifiers and integrate relevant functionalities into your biospecimen management system. *Biopreservation and Biobanking, 12*(3), 199–205.

36. Poloni, T. E., Medici, V., Carlos, A. F., Davin, A., Ceretti, A., Mangieri, M., Cassini, P., Vaccaro, R., Zaccaria, D., Abbondanza, S., Bordoni, M., Fantini, V., Fogato, E., Cereda, C., Ceroni, M., & Guaita, A. (2020). Abbiategrasso brain bank protocol for collecting, processing and characterizing aging brains. *Journal of Visualized Experiments: JoVE, 160*. https://doi.org/10.3791/60296

37. Vonsattel, J. P., Del Amaya, M. P., & Keller, C. E. (2008). Twenty-first century brain banking. Processing brains for research: The Columbia University methods. *Acta Neuropathologica, 115*(5), 509–532. https://doi.org/10.1007/s00401-007-0311-9

The Importance of Biobanking COVID-19 Samples

24

Araz Chiloyan and Lena Nanushyan

Abstract

Collection and storage of biological specimens has been a longstanding method of conducting research all around the world. Biobanking has allowed for all types of viruses to be studied by a global network of researchers and thus enhanced and accelerated the research and development of diagnostic tests and vaccines. However, with the introduction of the novel coronavirus (COVID-19) leading to a pandemic, the challenges and gaps in biobanking became more apparent as the entire world struggled to address the pandemic. With strict measures in place to reduce transmission of the disease, the idea of virtual biobanking began to take root as a method for researchers to access data irrespective of their location. With a better understanding of the gaps and challenges faced by researchers in biobanking and virtual biobanking, it is likely that there will be a stronger and more robust response to epidemics and pandemics in the future.

Keywords

Biobank · Coronavirus · Virus · Mutations · Virtual

Introduction

Biobanking specimens, biological tissue and health data, can provide insight into the genetic material of diseases and can play a crucial role for research in many different fields, specifically in healthcare [1]. When new viruses are discovered that have the potential to cause an epidemic or pandemic, there is a need for immediate response, which includes testing, surveillance and diagnostics [2]. In addition to the immediate

A. Chiloyan (✉) · L. Nanushyan
The Ministry of Health of the Republic of Armenia, Yerevan, Armenia
e-mail: chiloyan@bu.edu

emergency response, there is also a need for long-term research into understanding the new virus [2]. In order to understand the new virus, it is critical to have biobanks in place to allow for researchers to study the virus. With the discovery of SARS-CoV-2 (COVID-19) in 2019 and its continued presence in the world today, it is evident that biobanking allowed a network of global researchers to collaboratively study the fundamentals of the virus as well as its mutations and to expedite the development of a vaccine.

COVID-19 and Biobanking

SARS-Cov-2 is a highly transmissible virus that has a high mortality risk to humans. The symptoms vary from mild to severe with more serious symptoms consisting of trouble of breathing, chest pain or heaviness and loss of speech or movement. Due to the magnitude of those being affected, biobanking became a way to share information between institutions and countries to develop new diagnostic techniques to tackle the novel virus. Biobanking plays a crucial role in COVID-19 research with interim policies and guidelines for collection, processing, storage and shipment being developed on place—on demand in a very short time [3]. The guidelines that were set into place allowed for safe and efficient collection of samples to be used by institutions across the world, accelerating research and development.

Challenges of COVID-19 Biobanking

The demand for COVID-19 biobanking increased exponentially with the spread of the virus across the globe. With demand came challenges related to COVID-19 handling, operations, infrastructure support/resources and safety [4]. COVID-19 handling concerns included personal safety and guidelines due to the evolving knowledge about the virus [4, 5]. Operations and infrastructure support/resource concerns included sudden changes in routine activity, availability of competent personnel, and remote biobank operations with the new need for remote IT infrastructure [4]. Although many biobanking tools have been put into place to ensure proper management during crises, many biobanks were unprepared to address pandemic-related challenges that were long term [5, 6].

Virtual Biobanking for Future COVID-19 Research

The outbreak of the novel coronavirus disrupted clinical research functions worldwide as individuals were forced to practice social distancing in order to reduce the transmission of the virus. As the world shifted to a "work-from-home" mode, the idea of a "virtual" biobank, so-called "electronically simulated" biobank—a database of samples and associated data—was pushed towards the forefront as a potential vital resource to accelerate COVID-19 research [7, 8]. Global COVID-19 research would be advanced with the use of virtual biobanks, allowing researchers to source material and data irrespective of the location of the specimen [8]. There are currently many virtual biobanks worldwide including National Cancer Institute's Specimen Resource Locator, UK Biobank, EuroBioBank, Specimen Central, the UCL Virtual Biobank, and CHRISTUS Virtual Biobank (CVB) [8]. Setting up a virtual biobank requires financial, ethical and regulatory considerations. Safeguarding patient information, ownership of samples, determining infrastructure and necessary resources for forming a virtual COVID-19 biobank are key for its success [8].

Conclusion

Viral biobanks make it possible to archive samples of the virus and its mutated strains and permit a prompter and prepared response to future epidemics and pandemics [2]. COVID-19 biobanking allowed for acquisition and sharing of knowledge as quick as possible allowing for diagnostic strategies and vaccine development. However, there were many challenges associated with COVID-19 biobanking due to the long-term nature of the pandemic. Looking to the future, it is possible that with the set-up of COVID-19 virtual biobanking the world will be able to address concerns related to mutations in a faster and more collaborative way.

References

1. Hawkins, A. K. (2010). Biobanks: Importance, implications and opportunities for genetic counselors. *Journal of Genetic Counseling, 19*(5), 423–429. https://doi.org/10.1007/s10897-010-9305-1
2. EVA Zika Workgroup, Baronti, C., Lieutaud, P., Bardsley, M., de Lamballerie, X., Resman Rus, K., Korva, M., Petrovec, M., Avsic-Zupanc, T., Matusali, G., Meschi, S., EVA COVID-19 Workgroup, Baronti, C., Bardsley, M., Lieutaud, P., de Lamballerie, X., Blecker, T., Drexler, F., Drosten, C., Prat, C.M.A. (2020). The importance of biobanking for response to pandemics caused by emerging viruses: The European virus archive as an observatory of the global response to the zika virus and COVID-19 crisis. *Biopreservation and Biobanking, 18*(6), 569. https://doi.org/10.1089/bio.2020.0119
3. Gao, F., Tao, L., Ma, X., Lewandowski, D., & Shu, Z. (2020). A study of policies and guidelines for collecting, processing, and storing coronavirus disease 2019 patient biospecimens for biobanking and research. *Biopreservation and Biobanking, 18*(6), 511–516. https://doi.org/10.1089/bio.2020.0099
4. Allocca, C. M., Bledsoe, M. J., Albert, M., Anisimov, S. V., Bravo, E., Castelhano, M. G., Cohen, Y., De Wilde, M., Furuta, K., Kozlakidis, Z., Martin, D., Martins, A., McCall, S., Morrin, H., Pugh, R. S., Schacter, B., Simeon-Dubach, D., & Snapes, E. (2020). Biobanking in the COVID-19 era and beyond: Part 1. How early experiences can translate into actionable wisdom. *Biopreservation and Biobanking, 18*(6), 533–546. https://doi.org/10.1089/bio.2020.0082

5. Simeon-Dubach, D., & Henderson, M. K. (2020). Opportunities and risks for research biobanks in the COVID-19 era and beyond. *Biopreservation and Biobanking, 18*(6), 503–510. https://doi.org/10.1089/bio.2020.0079
6. Allocca, C. M., Snapes, E., Albert, M., Bledsoe, M. J., Castelhano, M. G., De Wilde, M., Furuta, K., Kozlakidis, Z., Martin, D., Martins, A., McCall, S. J., & Schacter, B. (2020). Biobanking in the COVID-19 era and beyond: Part 2. A set of tool implementation case studies. *Biopreservation and Biobanking, 18*(6), 547–560. https://doi.org/10.1089/bio.2020.0083
7. van Draanen, J., Davidson, P., Bour-Jordan, H., et al. (2017). Assessing researcher needs for a virtual biobank. *Biopreservation and Biobanking, 15*(3), 203–210. https://doi.org/10.1089/bio.2016.0009
8. Paul, S., & Chatterjee, M. K. (2020). Data sharing solutions for biobanks for the COVID-19 pandemic. *Biopreservation and Biobanking, 18*(6), 581–586. https://doi.org/10.1089/bio.2020.0040

Oncobiobanking in a Middle-Income Country: The Example of Russia

Andrey D. Kaprin, S. A. Ivanov, V. A. Petrov, L. J. Grivtsova, and Karine Sargsyan

Abstract

Biosamples and biological specimen banks are essential elements in the development of global cancer science and practice. Modern evidence-based medicine dictates the need to comply with fairly strict conditions by conducting scientific research. Taking into account the scale of the tasks solved by modern oncological research practice, it is necessary to create professional biobanks of samples from cancer patients and patients with borderline tumour conditions. The main task of such an infrastructure is the professional collection of biological samples that can be studied in relation to a wide range of molecular biological parameters and will not lose their information value even over a long storage period (from 10 to 100 years).

Keywords

Pathological tissue · Russian Federation · Oncobiobank · Epidermal growth factor receptor (EGFR) · OnkoBioBank · European Prospective Investigation into Cancer and Nutrition (EPIC) · Chernobyl tissue bank · Russian Biobank · National Association of Biobanks and Biobanking Specialists (NASBIO)

A. D. Kaprin (✉) · S. A. Ivanov · V. A. Petrov · L. J. Grivtsova
FGBY/FSBI (Federal state budgetary institution) "National Medical Research Center of Radiology" of the Ministry of Health of the Russian Federation, Moscow, Russia

K. Sargsyan
International Biobanking and Education, Medical University of Graz, Graz, Austria

Department of Medical Genetics, Yerevan State Medical University, Yerevan, Armenia

Ministry of Health of the Republic of Armenia, Yerevan, Armenia

175

Introduction

Biosamples and biological specimen banks are essential elements in the development of global cancer science and practice. Modern evidence-based medicine dictates the need to comply with fairly strict conditions by conducting scientific research. Taking into account the scale of the tasks solved by modern oncological research practice, it is necessary to create professional biobanks of samples from cancer patients and patients with borderline tumour conditions.

The main task of such a structure is the specialised collection of human biological material that are open to be investigated for a comprehensive range of medical parameters and will valuable in their scientific value over a long storing period (from 10 to 100 years).

A number of conditions must be met for a biobank to function successfully. There must be a concrete formulation to create an adequate infrastructure, including preanalytical handling at the stage of taking biological samples and their competent placement in the repository (including adequate labelling) and, finally, detailed information support (the so-called sample annotation), as well as a wide availability of collections of biological material for research. An extract from the sample (in the case of medical biobanks) includes non-personal data such as clinical and physical information, data of all examinations and changes in the patient's condition during the entire period of his/her observation, including medical and diagnostic manipulations, subject to the conditions of complete anonymity of the sample. It is important that even small-scale research projects carried out on biosamples of the repository (small sample number, individual molecular determinants) enrich the biocollection by supplementing the annotation of the stored reserve sample with new data (back up). This raises the challenge of appropriately accommodating and augmenting large data sets and ensuring optimal access to information. A condition for optimal functioning of a biobank infrastructure is clear regulatory legal information, a key element of which is informed consent for the voluntary donation of samples to the biobank, which allows the widest possible use of samples in scientific research.

Here, we describe some of the current challenges in biobanking as well as describe and analyse the currently existing infrastructure of oncological biobanks, both in the Russian Federation and worldwide.

Biobanks in General

Following the definition of ISBER, a biobank is an officially formed, physically existing, or virtual institution that can receive, process, store and distribute samples and associated data for the current use or future purposes (https://www.isber.org/). The objects of biobanking can be biological fluids (blood serum, saliva, urine), healthy and pathological tissues, cell cultures, strains of bacteria, viruses, physiological and pathological secretions, smears, scrapings, swabs, biopsies and surgical material (https://biospecimens.cancer.gov/patientcorner).

The value of biobanks, first of all, as a unique research resource and also as some kind of biological life insurance (and not only human), is currently beyond doubt. The topic of creating large-scale collections is not new to the world practice and is gaining popularity also in the Russian Federation [1].

Bioresource collections with an oncological focus were created and are being created practically in almost every planned scientific research, but, unfortunately, not all of them are correctly annotated, are not preserved for a long time or are consumed without remains. Accordingly, such bioresource collections do not grow to the level of a biobank. Thus, valuable scientific material is lost, and the scientific information obtained is often not properly promoted. Perhaps, this is because of the lack of consolidation of scientific knowledge. Despite the success in the diagnosis and treatment of cancer patients, the incidence of cancer in the Russian Federation is constantly growing and these countries are still far from defeating cancer [2]. That is why, in the era of targeted therapy and biotherapy, from the point of view of evidence-based medicine, it is relevant to create large-scale biobanks with an oncological focus.

How to Define a Biobank

The idea of a biobank is not new, collections of biomaterials have always been created, but not all of them can be considered a biobank. Practically in all pathomorphological departments of large hospitals and medical centres, as well as in clinical diagnostic laboratories, a certain number of biological samples of patients are stored (as a rule, blood serum, paraffin blocks, cytological preparations), but these collections cannot be considered a biobank due to several reasons. These specimens are limited in their ability to be studied, as they are taken from a specific patient for a specific purpose, and cannot be used for other research purposes. In addition, as a rule, pathology departments, cytology and other diagnostic laboratories do not have enough space to store a large number of samples of different types. This was a prerequisite for the formation of a new, special field of activity in medicine and biology, designated as biobanking, and the structure itself (the storage of biosamples) was called a biobank.

Several years ago, the opinion was voiced that a biobank is a repository of human biological samples used for scientific and medical purposes, while biorepositories are collections of various biological resources [3]. In our view, both terms mean the same thing—the adequate preservation of different types of biological specimens and the information associated with them as comprehensively as possible, regardless of whether or not they belong to a particular species.

Biobank activities may include not only the storage of specimens, combined with clinical, epidemiological and general background data, but also the biobank itself may perform some research functions, thereby increasing the scientific value of biocollections. It is important that specimen and annotation information, while respecting the anonymity of the donor's identity, should be made available to the wider scientific community, which will facilitate a wide variety of research on the collections. Properly organised and functioning biobanks also provide detailed information on the acquisition, processing and storage of each sample—e.g. time and method of collection, delivery circumstances [4].

Of equal importance to a well-functioning biobank is that all manipulations are carried out in accordance with SOPs governing the work of the unit at all stages of sample collection and processing, including transport features, preanalytical phase, analytics, anonymity of samples in the biobank database, storage and exchange conditions and timing and both of the samples themselves and of the data annotating them.

Thus, a biobank is determined not only by the number of samples stored in it, but also by all the accompanying information, as well as by a properly functioning infrastructure. It seems as if a biobank should be an independent subdivision with its own preanalytical and analytical unit, staff including IT specialists. Interestingly, the first attempt to create an independent unit with the goal of safekeeping of samples was made in 1948 as part of the Framingham Heart Study, a project to identify risk factors for cardiovascular disease based on a study of peripheral blood samples. It should be noted that the research results of this bank were published almost 20 years after the start of its formation [5].

For the successful functioning of the infrastructure, educational work with medical personnel is necessary, since medical doctors do not have a correct understanding of the importance of a biobank both for science and for practical medicine. A further important component of a successful biobank is to educate the public, explaining the aims and objectives of a biobank and its importance to the physical health of future generations.

General Principles of the Biobank Infrastructure

In summarising, the long-term experience of a number of institutions and projects involving a biobank resource, a biobank should consist of the following components: collection and storage of human samples in combination with therapeutic and epidemiological information; dynamic development of the biobank—continuous sample collection in the long term; connection of the biobank with ongoing research projects; respect for donor (sample donor/patient) anonymity and use of uniform standards and management procedures [6–12].

The process of establishing a biobank begins with defining the purpose of the project; the type of samples is also defined at this stage. The process of obtaining a sample begins with obtaining the informed consent of the patient, and this document must be accepted and permitted by the local ethics committee.

The collection of blood samples and other fluids for biobanking is usually carried out at the primary diagnostic stage and is possible in both outpatient and 24-h inpatient settings. If tissue is collected from a patient with, e.g. cancer (biopsy material, surgical material) for research purposes, the need and importance of a complete primary diagnosis should be considered in the first instance. Tissue samples from cancer patients are only placed in a biobank when the necessary amount of material for a complete diagnosis and staging of the tumour has been obtained. At this stage, the involvement of the pathology service is essential. The involvement of a pathologist at the biobanking stage, in addition to avoiding interference in the diagnostic process, also makes it possible to confirm the quality of the sample (absence of undesirable artefacts, necrosis, etc.).

After collection of any biosample, with documentation of all transportation conditions (time, temperature, availability of a directional description), the sample goes to the biobank, where it is anonymised/pseudonymised, marked and processed. At this stage, the crucial step is aliquoting the samples. Aliquoting is the process of dividing either a native sample or an isolated fraction (plasma, serum) into smaller samples to create backup copies of it, allowing a sample of the same patient to be examined several times. For example, in the UK Biobank, in accordance with the adopted SOP, 19 aliquots are prepared from five primary samples of urine and blood of a patient [9]. Upon completion of aliquoting, the samples are bar-coded and stored, with registration in the electronic database of the biobank.

A very important aspect is the choice of containers for placing the aliquots. Placing samples in containers that are not intended for freezing is unacceptable and will negatively influence the value of the stored samples. The quality of polypropylene, from which the cryocontainers are made, is also important [13].

Another key point of the preanalytical phase is the implementation of an exact reproducible aliquoting process, which is usually carried out manually due to cost savings. However, automation of the aliquoting process seems to be the most appropriate way, which reduces the number of samples required, eliminates human error and facilitates the labelling and subsequent identification of the samples.

Considering the multi-step process, all steps of the process must be easily reproducible, which requires clear regulations.

The functions of the networking bodies for biobanks are currently fulfilled by international organisations such as ISBER and ESBB. ISBER's functions are broad—developing biobanking and an international network of biobanks, accumulating international experience concerning all aspects of biobanking and its stages and, as mentioned above, developing recommendations. As of the end of 2019, three Russian organisations are members of ISBER, and one is member of ESBB. The "Guidelines for standardised biobanking" published in 2010 by Irish molecular medicine specialists [14] are also among the useful sources of information. The International Organisation for Standardisation ISO published the ISO20387:2018 biobanking standard in August 2018 [15]. The document includes general requirements for biobanking organisations. ISO 20387:2018 was based on ISO9001.

The Tasks of an Oncobiobank

Primarily, a biobank is a ready-made source of high-quality (tissue) samples. Biosamples can contain certain genomic, epigenomic, transcriptomic, proteomic and metabolic changes that characterise a specific nosological form of cancer in conjunction with the patient's personal/individual characteristics. This in itself is extremely important data, which can be analysed in correlation with the clinical information of the patient to determine the impact of molecular biological features on the efficacy of the therapy provided. In doing so, the biobank, by providing a sufficient amount of study material, provides the study with weighty statistical significance. With this in mind, the biobank should have a robust computational/analytical and information infrastructure to manage the data and conduct meaningful analyses.

Another point is the possibility of finding new predictors and drug targets, as quite often molecularly targeted therapies achieve long-term cancer control, but 15–20% of patients develop resistance to this type of treatment, which leads to the progression of the disease.

A very important point for the biobank, and especially for the oncological biobank, is the preservation of the original samples (conditional point of return, the so-called back up). Over time, the research capabilities change, and if, after a lapse of time from the diagnosis (after 3–5–10 years), we can return to the original tumour and study it in more detail, already knowing what happened to the patient during this period. This will undoubtedly bring us closer to solving the problem of a particular type of cancer.

Using molecular-targeted therapy, we may change the biological portrait of the tumour, possibly "forcing" it to mutate. However, subject to the formation of biobanks, it is possible to return to the initial data, and not only for research purposes, but also for the development of personalized therapy, taking into account the characteristics of the primary tumour [16].

Therefore, in recent years, publications [17] have appeared which speak of high tumour heterogeneity and variability in the treatment process, which in turn poses the problem of the need to develop optimal methods of radiation therapy.

Research in Oncology with the Involvement of Biobank Resources

As noted above, biobanks are an essential tool for many tasks in oncology, as the search for adequate predictors of cancer risk and response to therapy is still an urgent task.

The EPIC (European Prospective Investigation into Cancer and Nutrition) biobank project is one of the most successful oncology biobank projects. Several large-scale studies have been carried out on samples concentrated in biobanks as part of this project. For example, an association between insulin-like growth factor 1 (IGF-1) and the risk of invasive ovarian cancer has been established [18]. The connection between lifestyle features and risk of breast cancer in menopausal women has been shown using biobank samples from this project [19].

Of particular note is a large-scale cohort study from 2014 to identify genetic predictors of prostate cancer mortality risk using the International Agency for Research on Cancer, Lyon, France Department of Biobank Research. This study included samples from 10,487 prostate cancer patients and 11,024 control individuals and analysed the association of 47 single-nucleotide polymorphisms identified in prostate cancer with cancer mortality. Eight variants of single-nucleotide polymorphisms were found associated with the risk of prostate cancer mortality. The study found that only one of the studied single-chain polymorphisms, rs11672691, is a factor of poor prognosis. Patients with this determinant had a shorter life expectancy. In contrast, detection of the other seven single-stranded polymorphisms (rs13385191 [C2orf43], rs17021918 [PDLIM5], rs10486567 [JAZF1], rs6465657 [LMTK2], rs7127900 (intergenic), rs2735839 [KLK3], rs10993994 [MSMB], rs13385191 [C2orf43]) determined longer patient survival [20].

A further larger study (involving 140,000 patients with retinitis pigmentosa (RP)) using the resources of several biobanks, has identified 63 new loci allied alongside the likelihood of prostate cancer, including inherited forms of the disease [21].

Another example of realizing the potential of biobanks is the investigation of risk factors for the development of ductal carcinoma of the mammary gland in situ, a condition, although not obligatory, but preceding the development of invasive breast cancer [22].

A study was carried out on samples from 263,788 women aged 40 to 69 years held in the UK Biobank population-based cohort repository. The study collected data on demographic, reproductive and medical factors, using computerised questionnaires, and ascertained the presence of ductal carcinoma of the breast in situ, using data from the UK Cancer Registry [23]. A number of factors were found to be related with the risk of developing ductal carcinoma in situ, using multivariate analysis. The most important result of this largest epidemiological study to date has been the identification of a particular risk group for this condition: postmenopausal women who are not on hormone replacement therapy and have an elevated body mass index [23]. The studies presented above, using biobank resources, were nosologically oriented.

It should be clear that the collection and database of the same biobank may serve very different purposes. The Australian scientists' prospective cohort study is an example of this. Initially, when the sampling started and databases were formed (1994–1995), the study aimed at analysing the relationship of characteristics of metabolism and the risk of cardiovascular disease [24, 25]. However, taking into consideration the peculiarities of the area where the population lives, the coastal zone (Busselton region) of Western Australia, which is considered to be an area with iodine deficiency [26], it was hypothesized that the thyroid hormone level can be associated with the risk of several types of cancer [27]. A relevant study was carried out, including 4843 individuals who were subjected to a very careful anamnestic questionnaire, anthropometric examination and a blood sample for the assessment of hormonal status in 1994/1995. The observation was carried out between 1994 and 2014. During the 20-year period, 16% of the study participants (600 people) were diagnosed with any cancer. 126 (8%) men and 100 (5%) women were spotted with diagnosis of prostate and breast cancer; correspondingly, 103 (3%) subjects developed colorectal cancer and 41 (1%) were diagnosed with lung cancer. No association was found between thyroid hormone levels and thyroperoxidase antibodies and the risk of developing breast cancer, colorectal cancer or lung cancer. However, the study showed a significant correlation between the development of prostate cancer and levels of thyrotropic hormone, TSH (inverse correlation) and thyroxine FT4 (direct correlation). A further vector of research has thus been determined to explore the function of the thyroid function in the increase of prostate cancer. This kind of research could not have been realised without the resources of biobanks.

All of the above confirm the importance of (onco)biobanks for obtaining clinically relevant information in oncology practice, both in terms of finding predictors of disease and improving screening programmes. Finally, the history of the introduction into clinical practice of the targeted drug gefitinib, initially targeting inactivation of the epidermal growth factor receptor (EGFR), is very illuminating. EGFR inhibitors were considered as universal agents, as an increase in the expression of these receptors is characteristic of most tumours of epithelial origin. Clinical trials of gefitinib showed very modest results, and the international clinical trial of ISEL was suspended in 2004. Samples from the patients included in the study were, however, retained. Summaries of the results of these studies, performed after the data of all participants had been collected, processed and archived, made it possible to perform nucleotide sequence analysis of the EGFR gene in tumours with and without treatment effect and to conclude that recurrent intragenic mutations are associated with drug efficacy. Thus, a drug initially designed to inactivate the normal EGFR gene was effective against the mutated gene [28].

The discovery of EGF receptor mutations and the following evidence of the effectiveness of EGF receptor tyrosine kinase inhibitors in targeting tumour genotype marked the age of precision medicine for non-small cell lung cancer. So both, the first as well as the second-generation EGF receptor tyrosine kinase inhibitors (erlotinib, gefitinib and afatinib), have been shown to be more effective than a platinum-based chemotherapy and represent the standard of care for patients with advanced non-small-cell lung cancer with an EGFR mutation. However, tumours can invariably develop resistance to first- and second-generation EGF receptor tyrosine kinase inhibitors as well, limiting their long-term efficacy. The most frequent mutations in the EGFR gene include deletions in exon 19 and a point mutation of L858R in exon 21. A secondary mutation in the EGFR gene, T790M, has been identified as the most common mutation responsible for the resistance to first- and second-generation inhibitors. Osimertinib, a third-generation inhibitor, blocks the EGF receptor tyrosine kinase both in the presence of activating mutations (del19 or L858R) and with the T790M mutation in the EGFR gene [29].

Thus, researchers have been faced with the need to develop tests for the detection of mutations in the EGFR gene to stratify patients and prescribe therapy depending on their presence [30]. The development and registration of a test is based on the identification of a predictive biomarker and requires analytical validation, which is carried out on biomaterial of different formats with clinical and laboratory information about the patient. In this context, the biobank is the main source of information for selecting specific groups of patients—carriers of mutations with or without treatment effect, determining whether the drug used will be effective. In addition, in terms of current research projects, cell suspensions isolated from a patient's tumour tissue sample can be a resource for obtaining an in vitro tumour model, PDC (Patient-Derived Cell), and an in vivo human tumour xenograft—PDX (Patient-Derived Xenograft) [31, 32].

In this way, the collections of primary tumour cells are created and organoid biorepositories—3D matrices, where tumour cells including tumour stem cells and tumour microenvironment cells including immune cells and stromal cells are co-cultured, are created if culturing conditions are optimised [33]. Such models appear to be the most promising and scientifically valid tools in the discovery and development of new generation drugs.

Veterinary cancer biobanks have also shown their relevance. A number of examples demonstrate the involvement of animals in preclinical studies, such as the first drug in the Bruton's tyrosine kinase inhibitor class, ibrutinib [34]. This is because existing experimental models of B-cell lymphoma in mice have proved insensitive to ibrutinib, and dogs with B-cell lymphoma are the only model to prove the mechanism of action of the drug in which the therapeutic effect of ibrutinib has been demonstrated, which provided a strong incentive to start clinical trials. Ibrutinib is currently a registered drug for the treatment of certain types of B-cell lymphoma.

In addition to demonstrating the enormous potential of a resource, such as an oncobobank, these studies also raise the question of the need for very careful bioinformatics analysis of the vast amounts of data generated by multi-component analysis of large numbers of samples.

Another important conclusion from these studies is the obviousness of combining the efforts of several biorepositories and the formation of so-called networks of oncobiobanks.

Unfortunately, there are almost no such examples of the implementation of large-scale cohort studies in oncology in the Russian Federation. It seems that the right collections of biological samples are still being formed.

The Russian Biobank Landscape

In Russia, the National Association of Biobanks and Biobanking Specialists (NASBIO, nasbio.ru) was established in 2018 with the support of the Ministry of Health of the Russian Federation, of which FGBU NRRC Radiology of the Russian Ministry of Health is also a member. Its primary goal is to consolidate efforts to form a full-fledged infrastructure of biobanks, depositories and collections of biomaterials in the country [35].

At present, the Federal Agency for Technical Regulation and Metrology has begun translating ISO20387:2018 into Russian, as a result of the NASBIO initiative.

At the moment, biological material banks, DNA depositories, stem cell banks, donor material banks, repositories, etc. are being established in the Russian Federation. Collections with samples of seeds, plant tissue cultures, animal DNA, microorganisms and viruses are being formed. For example, in 2010 a regional genetic bank of rare and endangered plant species included in the Red Book of Volgograd Oblast was created. By the end of 2020, there were between 20 and

30 biobanks in the Russian Federation, as well as more than 200 collections of biological samples [36]. An interesting Russian project is the marine biobank (Center for Shared Use of the Resource Collection "Marine Biobank"). The Marine Biobank was organized in 2017 based on the National Scientific Center for Marine Biology of the Far Eastern Branch of the Russian Academy of Sciences. The goal of the Marine Biobank is to ensure the conduct of scientific research using the existing collections and equipment in accordance with international protocols for the collection, cataloguing, maintenance and storage of biological samples of marine origin. The collection currently consists of over 200,000 items and includes both living collections of marine microorganisms and the stock collections of the Museum and Marine Biology Research Centre. The collections are represented by littoral, shelf and deep-water collections from the Far East seas of Russia, with a significant part of the specimens collected in the coastal waters of Vietnam and other areas of the World Ocean.

In terms of the number of human tissue samples stored, the leaders are the biobanks of the Academician V.I. Kulakov Scientific Research Centre for Obstetrics, Gynecology and Perinatology, the N.F. Gamaleya Research Centre for Epidemiology and Microbiology, the N.F. Gamaleya Research Centre for Epidemiology and Microbiology and the St. Petersburg State University Biobank (about 200,000 samples are stored), N.F. Gamaleya, Biobank Centre of St. Petersburg State University (about 200,000 samples are stored). However, the information on these collections is not sufficiently available even on the websites of the organisations.

The issues and problems in the use of pathology specimens should be highlighted. Currently, every pathology department in the country and in the world stores paraffin-embedded tissue specimens (paraffin blocks) and histological specimens (slides) for at least 25 years after collecting the samples. There are tens of millions of specimens stored in pathology archives around the country. Digital archives of pathology specimens are virtually non-existent in Russia, as is tissue microarray technology. The use of collected pathology specimens for scientific purposes is hampered by the lack of appropriate informed consent and associated information. The involvement of pathology archives in biobanking activities is thus very low.

Oncobiobank of the National Research Institute of Radiology of the Russian Ministry of Health

Analysing the experience of the most successful Russian and world biobanks and biorepositories, at the end of 2018, the formation of a biobank of samples of cancer patients (hereinafter—the Centre) was started on the basis of FGBU "NRC Radiology" of the Ministry of Health of Russia (hereinafter—OnkoBioBank, https://mrrc.nmicr.ru/filialy/laboratoriya-klinicheskoy-immunologii2/). This oncobiobank is being created as a long-term infrastructure for the storage of samples and data array with the possibility of research of biological material, both in the near future and in the future. The main goal is to create a solid foundation for challenges in the fields of oncology, biotechnology and genetic engineering.

In May 2019, at the first scientific and practical conference "Molecular Fundamentals of Clinical Oncology", held at the A.F. Tsyb MRRC, a branch of the Federal State Budgetary Institution "NRC Radiology" (hereinafter, Obninsk branch), a round table dedicated to biobanks was held, where a proposal was first voiced about the possibility of creating a network biobank of samples of cancer patients of the Russian Federation (hereinafter, Network Biobank) on the basis of the Centre. Oncological institutions of the North-West and Central Federal Districts of the Russian Federation were invited to create the Network Biobank. Consent for cooperation has been received from 17 subjects of the Russian Federation. The decision to establish a network biobank of samples of oncology patients has been supported by the Ministry of Health of the Russian Federation.

To date, OnkoBioBank is a structural subdivision of the Laboratory Medicine Department, which is represented by clinical, biochemical and immunological laboratories of high level of equipment, both from the technical side and in terms of specialists (see Fig. 25.1).

An electronic case history is maintained and archived in a single system, which integrates the OnkoBioBank biosamples registration and recording system, which allows for maximum annotation of biosamples. OnkoBioBank's operations are regulated according to international standards, respecting all confidentiality norms with regard to the preservation of personal information. Standard operating procedures, including preanalytics, analytics and sample annotation, are developed based on international protocols with the involvement of expert lawyers and an ethics committee, and a single informed consent has been developed. The possibility of a cloud-based data repository is being explored as being the most economically and practically optimal [37, 38], especially in the framework of the Network Biobank. This will enable the efficient use of information in clinical and scientific research for a larger number of researchers.

The OncoBioBank is a multifunctional structure (see Fig. 25.2).

Since the Centre is licensed and performs hematopoietic stem cell transplants, a cryobank of this transplant material is available. Since the late 1990s, a bone marrow mesenchymal stromal/stem cells (MSCs) bank has been established and operated at the Obninsk branch; its collection currently consists of 773 aliquots of undifferentiated human and animal (rat, elk) bone marrow MSCs. MSC cultures are used in the development of cell therapy programmes, in particular to prevent anthracycline cardiomyopathy in cancer patients [39]. The fertility bank is actively functioning, where for more than 10 years, ovarian tissue and oocytes of cancer patients have been preserved to maintain fertility after chemo- and hormone therapy.

At the end of 2020, the Centre initiated a tumour biobanking programme for pets. The collection is small and consists of 30 specimens including 17 spontaneous canine tumours of different histogenesis, 6 benign canine tumours and 7 spontaneous feline tumours. The establishment of a pet tumour biobank, a veterinary biorepository, is an infrastructural link in a chain of major research work, including basic, preclinical, translational and clinical phases. A large collection of immortalised human and animal tumour cell lines has been assembled at the Centre over many years of work. The collection includes more than 40 cell lines obtained

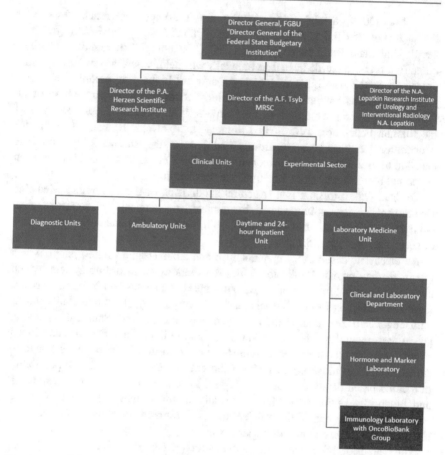

Fig. 25.1 Laboratory Medicine Department

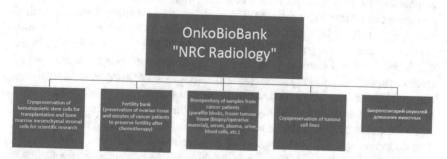

Fig. 25.2 OnkoBioBank "NRC Radiology"

from accredited international (ATCC, ECACC) and Russian repositories (Institute of Virology, Research Institute of Cytology of the Russian Academy of Sciences, Cancer Research Centre, Federal State Budgetary Educational Institution of the

Russian Academy of Sciences, the Russian Academy of Medical Sciences, the Russian Academy of Sciences and the Russian Academy of Sciences. Scientific Research Center for Oncology of the Ministry of Health of Russia). Immortalized human tumour cells include carcinoma, epidermoid carcinoma, squamous cell carcinoma, adenocarcinoma, transitional cell carcinoma, glioblastoma and fibrosarcoma. Cell cultures of immortalized mouse tumour cells include sarcoma, carcinoma, adenocarcinoma, lymphocytic leukaemia, myelomonocytic leukaemia, melanoma and leukaemia.

The resource of this biorepository was the basis for a wide range of experimental studies to identify potential anti-tumour activities and rapid identification of new therapeutic agents, including chemotherapeutic, targeting, radiopharmaceuticals, photosensitizers for photodynamic therapy and gene therapy drugs [40–54].

The biorepository of biological samples from cancer patients at the Centre is not very large, comprising 2850 samples from 550 cancer patients. As part of the projects was implemented on the basis of the OnkoBioBank, the main nosological forms such as lung, stomach, oesophagus, colorectal cancer, ovarian cancer, prostate cancer, breast cancer, tumours of the lymphoid and hematopoietic system are currently being collected. With the involvement of the Centre's bioresource collection for studying the effectiveness of neoadjuvant therapy in patients suffering from gastric cancer, a method of immunological evaluation of tumour micro metastases to signal lymph nodes has been developed. An invention patent was obtained (RU2727251 21.07.2020).

It should be noted that the Obninsk Branch is a participant of the international network biobank "Chernobyl Tissue Bank", including collections of Russia, Ukraine and Belarus; within the framework of this project, blood components and tissue components of the tumours of 1250 patients are stored. Thus, a multifunctional structure of OncoBioBank has been formed based on the Centre, covering the maximum possible aspects of this process in oncological science and practice, including biorepositories of cultures of tumour cells, malignant tumours of humans and animals and collections of various types of biological samples from cancer patients.

Biobanking in the Future

Biobanking is developing worldwide, but there are no universally acknowledged industry standards or worldwide registries yet. Creating standardised protocols should significantly improve the efficacy of clinical data collection, as different institutions use identical methods to characterise and annotate samples stored in their respective biobanks. Standardisation can be greatly accelerated by automated systems that organise sample collections into predefined categories based on tissue type, molecular marker or preservation method. This would greatly improve and simplify researchers' access to biospecimen collections, allowing more research to be conducted and increasing the value of biological specimen collections.

Unfortunately, most biobanks currently operate completely independently of each other, each according to their own individual standards for sample organisation and data collection. This makes collaboration between biobanks difficult and requires further methods to merge the diverse data collections into a solitary registry for data comparison or aggregation. Such a global registry encompassing all biobanks seems to be a favourable future for biobanking, demarcated by comprehensive metadata and broad partnership to promote medical research.

References

1. Maleina, M. N. (2020). Legal status of biobank (bank of human biological materials). *Law. Journal of the Higher School of Economics, 1*, 98–117. https://doi.org/10.17323/2072-8166.2020.1.98.117
2. Malignant neoplasms in Russia in 2018 (morbidity and mortality) Ed. HELL. Kaprina, V.V. Starinsky, G.V. Petrova - M .: MNIOI im. P.A. Herzen – branch of the Federal State Budgetary Institution "National Medical Research Center of Radiology" of the Ministry of Health of Russia, 2019.250 p. ISBN 978-5-85502-251-3 . *Radiation and risk, 26*(2), 26–37. https://doi.org/10.21870/0131-3878-2017-26-2-26-40
3. Smirnova, J. (2013). Banking as a path to personalized medicine. *Science and Life, 1*, 14s.
4. Fransson, M. N., Rial-Sebbag, E., Brochhausen, M., & Litton, J. E. (2015). Toward a common language for biobanking. *European Journal of Human Genetics, 23*(1), 22–28. https://doi.org/10.1038/ejhg.2014.45
5. Riegman, P. H., Morente, M. M., Betsou, F., de Blasio, P., & Geary, P. (2008). Biobanking for better healthcare. *Molecular Oncology, 2*(3), 213–222. https://doi.org/10.1016/j.molonc.2008.07.004
6. Dawber, T. R., & Kannel, W. B. (1966). The Framingham study. An epidemiologic approach to coronary heart disease. *Circulation, 34*(4), 553–555. https://doi.org/10.1161/01.cir.34.4.553
7. Asslaber, M., & Zatloukal, K. (2007). Biobanks: Transnational, European and global networks. *Briefings in Functional Genomics and Proteomics, 6*(3), 193–201. https://doi.org/10.1093/bfgp/elm023
8. Swede, H., Stone, C. L., & Norwood, A. R. (2007). National population-based biobanks for genetic research. *Genetics in Medicine, 9*(3), 141–149. https://doi.org/10.1097/gim.0b013e3180330039
9. Mohamadkhani, A., & Poustchi, H. (2015). Repository of human blood derivative biospecimens in biobank: Technical implications. *Middle East Journal of Digestive Diseases, 7*(2), 61–66.
10. Andersson, K., Bray, F., Arbyn, M., Storm, H., Zanetti, R., Hallmans, G., Coegergh, J. W., & Dillner, J. (2010). The interface of population-based cancer registries and biobanks in etiological and clinical research–current and future perspectives. *Acta Oncologica, 49*(8), 1227–1234. https://doi.org/10.3109/0284186X.2010.496792
11. Elliott, P., Peakman, T. C., & Biobank, U. K. (2008). The UK Biobank sample handling and storage protocol for the collection, processing and archiving of human blood and urine. *International Journal of Epidemiology, 37*(2), 234–244. https://doi.org/10.1093/ije/dym276
12. Paskal, W., Paskal, A. M., Dębski, T., Gryziak, M., & Jaworowski, J. (2018). Aspects of modern biobank activity – Comprehensive review. *Pathology & Oncology Research, 24*, 771–785. https://doi.org/10.1007/s12253-018-0418-4
13. Kofanova, O. A., Mommaerts, K., & Betsou, F. (2015). Tube polypropylene: A neglected critical parameter for protein adsorption during biospecimen storage. *Biopreservation and Biobanking, 13*(4), 296–298. https://doi.org/10.1089/bio.2014.0082

14. Guerin, J. S., Murray, D. W., McGrath, M. M., Yuille, M. A., McPartlin, J. M., & Doran, P. P. (2010). Molecular medicine Ireland guidelines for standardized biobanking. *Biopreservation and Biobanking, 8*(1), 3–63. https://doi.org/10.1089/bio.2010.8101

15. ISO 20387:18 Biotechnology – Biobanking – General Requirements for Biobanking. (2018). Accessed March 08, 2021, from http://www.iso.org/standard/67888.html

16. Anisimov, S. V., Meshkov, A. N., Glotov, A. S., Borisova, A. L., Balanovsky, O. P., Belyaev, V. E., Granstrem, O. K., Grivtsova, L. Y., Efimenko, A. Y., Pokrovskaya, M. S., Semenenko, T. A., Sukhorukov, V. S., Kaprin, A. D., & Dapkina, O. M. (2020). National Association of biobanks and biobanking specialists: New community for promoting biobanking ideas and projects in Russia. *Biopreservation and Biobanking, 19*(1). https://doi.org/10.1089/bio.2020.0049

17. Caprin, A. D. (2020). The oncology system in Russia is developing at an unprecedented rate. Accessed 12.01.2020, from https://ria.ru/20190918/1558771870.html

18. Kaprin, A. D., Galkin, V. N., Zhavoronkov, L. P., Ivanov, V. K., Ivanov, S. A., & Romanski, Y. S. (2017). *Synthesis of basic and applied research is the basis of obtaining high-quality findings and translating them into clinical practice.*

19. McKenzie, F., Ferrari, P., Freisling, H., Chajés, V., Rinaldi, S., de Batlle, J., Dahm, C. C., Overvad, K., Baglietto, L., Dartois, L., Dossus, L., Lagiou, P., Trichopoulos, D., Trichopoulou, A., Krogh, V., Panico, S., Tumino, R., Rosso, S., Bueno-de-Mesquita, H. B., et al. (2014). Healthy lifestyle and risk of breast cancer among postmenopausal women in the European prospective investigation into cancer and nutrition cohort study. *International Journal of Cancer, 136*(11), 2640–2648. https://doi.org/10.1002/ijc.29315

20. Shui, I. M., Lindstrum, S., Kibel, A. S., Berndt, S. I., Campa, D., Gerke, T., Penney, K. L., Albanes, D., Berg, C., Bueno-de-Mesquita, H. B., Chanock, S., Crawford, E. D., Diver, W. R., Gapstur, S. M., Gaziano, J. M., Giles, G. G., Henderson, B., Hoover, R., Johansson, M., … Kraft, P. (2014). Prostate cancer (PCa) risk variants and risk of fatal PCa in the National Cancer Institute Breast and Prostate Cancer Cohort Consortium. *European Urology, 65*(6), 1069–1075. https://doi.org/10.1016/j.eururo.2013.12.058

21. Schumacher, F. R., Olama, A. A., Berndt, S. I., Benlloch, S., Ahmed, M., Saunders, E. J., Dadaev, T., Leongamornlert, D., Anokian, E., Cieza-Borrella, C., Goh, C., Brook, M. N., Sheng, X., Fachal, L., Dennis, J., Tyrer, J., Muir, K., Lophatananon, A., Stevens, V. L. The genetic associations and mechanisms in oncology (GAME-ON)/elucidating loci involved in prostate cancer susceptibility (ELLIPSE) consortium. Association analyses of more than 140,000 men identify 63 new prostate cancer susceptibility loci. *Nature Genetics, 50*(7), 928–936. https://doi.org/10.1038/s41588-018-0142-8

22. Erbas, B., Provenzano, E., Armes, J., & Gertig, D. (2006). The natural history of ductal carcinoma in situ of the breast: A review. *Breast Cancer Research and Treatment, 97*(2), 135–144. https://doi.org/10.1007/s10549-005-9101-z

23. Peila, R., Arthur, R., & Rohan, T. E. (2020). Risk factors for ductal carcinoma in situ of the breast in the UK Biobank cohort study. *Cancer Epidemiology, 64*(101648). https://doi.org/10.1016/j.canep.2019.101648

24. Knuiman, M. W., Jamrozik, K., Welborn, T. A., Bulsara, M. K., Divitini, M. L., & Whittall, D. E. (1995). Age and secular trends in risk factors for cardiovascular disease in Busselton. *Australian Journal of Public Health, 19*(4), 375–382. https://doi.org/10.1111/j.1753-6405.1995.tb00389.x

25. Knuiman, M. W., Hung, J., Divitini, M. L., Davis, T. M., & Beilby, J. P. (2009). Utility of the metabolic syndrome and its components in the prediction of incident cardiovascular disease: A prospective cohort study. *European Journal of Cardiovascular Prevention and Rehabilitation, 16*(2), 235–241. https://doi.org/10.1097/HJR.0b013e32832955fc

26. Li, M., Eastman, C. J., Waite, K. V., Ma, G., Zacharin, M. R., Topliss, D. J., Harding, P. E., Walsh, J. P., Lynley, C. W., Mortimer, R. H., Mackenzie, E. J., Byth, K., & Doyle, Z. (2006). Are Australian children iodine deficient? Results of the Australian National Iodine Nutrition Study. *Medical Journal of Australia, 184*(4), 165–169. https://doi.org/10.5694/j.1326-5377.2008.tb01831.x

27. Chan, Y. X., Knuiman, M. W., Divitini, M. L., Brown, S. J., Walsh, J., & Yeap, B. B. (2017). Lower TSH and higher free thyroxine predict incidence of prostate but not breast, colorectal or lung cancer. *European Journal of Endocrinology, 177*(4), 297–308. https://doi.org/10.1530/EJE-17-0197

28. Dickran, K., Blumenthal, G. M., Yuan, W., He, K., Keegan, P., & Pazdur, R. (2016). FDA approval of Gefitinib for the treatment of patients with metastatic EGFR mutation–positive non–small cell lung cancer. *Clinical Cancer Research, 22*(6), 1307–1312. https://doi.org/10.1158/1078-0432.CCR-15-2266

29. Santarpia, M., Liguori, A., Karachaliou, N., Gonzalez-Cao, M., Daffinà, M. G., D'Aveni, A., Marabello, G., & Rosell, R. (2017). Osimertinib in the treatment of non-small-cell lung cancer: Design, development and place in therapy. *Lung Cancer (Auckl), 2017*(8), 109–125. https://doi.org/10.2147/LCTT.S119644

30. Arbour, K. C., & Riely, G. J. (2019). Systemic therapy for locally advanced and metastatic non-small cell lung cancer: A review. *JAMA, 322*(8), 764–774. https://doi.org/10.1001/jama.2019.11058

31. Bolck, H. A., Pauli, C., Göbel, E., Mühlbauer, K., Dettwiler, S., Moch, H., & Schraml, P. (2019). Cancer sample biobanking at the next level: Combining tissue with living cell repositories to promote precision medicine. *Frontiers in Cell and Development Biology, 7*, 246. https://doi.org/10.3389/fcell.2019.00246

32. Annibali, D., Leucci, E., Hermans, E., & Amant, F. (2019). Development of patient-derived tumor xenograft models. *Methods in Molecular Biology, 1862*, 217–225. https://doi.org/10.1007/978-1-4939-8769-6_15

33. Osswald, A., Hedrich, V., & Sommergruber, W. (2019). 3D-3 tumor models in drug discovery for analysis of immune cell infiltration. *Methods in Molecular Biology, 1953*, 151–162. https://doi.org/10.1007/978-1-4939-9145-7_10

34. Honigberg, L. A., Smith, A. M., Sirisawad, M., Verner, E., Loury, D., Chang, B., Li, S., Pan, Z., Thamm, D. H., Miller, R. A., & Buggy, J. J. (2010). The Bruton tyrosine kinase inhibitor PCI-32765 blocks B-cell activation and is efficacious in models of autoimmune disease and B-cell malignancy. *PNAS, 107*(29), 13075–13080. https://doi.org/10.1073/pnas.1004594107

35. Im, K., Gui, D., & Yong, W. H. (2019). An introduction to hardware, software, and other information technology needs of biomedical biobanks. *Methods in Molecular Biology, 1897*, 17–29. https://doi.org/10.1007/978-1-4939-8935-5_3

36. Tumanov, R. (2019). How it works: "Biobanks" in Russia. Accessed February 02, 2019, from https://spbvedomosti.ru/news/country_and_world/donor_idet_v_bank

37. Ose, J., Fortner, R. T., Schock, H., Peeters, P. H., Onland-Moret, N. C., Bueno-de-Mesquita, H. B., Weiderpass, E., Gram, I. T., Overvad, K., Tjonneland, A., Dossus, L., Fournier, A., Baglietto, L., Trichopoulou, A., Benetou, V., Trichopoulos, D., Boeing, H., Masala, G., Krogh, V., & Kaaks, R. (2015). Insulin-like growth factor I and risk of epithelial invasive ovarian cancer by tumour characteristics: Results from the EPIC cohort. *British Journal of Cancer, 112*(1), 162–166. https://doi.org/10.1038/bjc.2014.566

38. Eydeler, K. (2013). Why cloud-based LIMS is ideal for biobanking. Accessed March 03, 2016, from http://www.shonan-village.co.jp/anrrc2013/pdf/S2-1.pdf

39. Grivtsova, L. Y., Popovkina, O. E., Dukhova, N. N., Politiko, O. A., Yuzhakov, V. V., Lepekhina, L. A., Kalsina, S. S., Ivanov, S. A., & Kaprin, A. (2020). Cell biobank as a necessary infrastructure for the development and implementation of mesenchymal stem cell-based therapy in the treatment of anthracycline-induced cardiotoxicity. Literature review and own data. *Cardiovascular Therapy and Prevention, 19*(6), 2733. https://doi.org/10.15829/1728-8800-2020-2733

40. Gerasimova, G. K., Yakubovskaya, R. I., Pankratov, A. A., Treshchalina, E. M., Nemtosova, E. R., Andreeva, T. N., Venediktova, Y. B., Plyutinskaya, A. D., Bezborodova, O. A., Sidorova, T. A., Baryshnikov, A. B., Kaliya, O. L., Voroztsov, G. N., & Luzhkov, Y. M. (2015). Binary catalytic therapy: A new approach to treatment of malignant tumors. Results of pre-clinical and

clinical studies. *Russian Journal of General Chemistry, 85*(1), 289–302. https://doi.org/10.1134/S1070363215010442

41. Nemtsova, E. P., Tikhonova, E. G., Bezborodova, O. A., Pankratov, A. A., Venediktova, J. B., Korotkevich, E. I., Kostryukova, L. V., & Tereshkina, J. A. (2020). Preclinical study of the pharmacological properties of doxorubicin-NPh. *Bulletin of Experimental Biology and Medicine., 169*(6), 720–726. https://doi.org/10.1007/s10517-020-04977-5

42. Nemtsova, E. R., Bezborodova, O. A., Morozova, N. B., Vorontsova, M. S., Venediktova, J. B., Andreeva, T. N., Nesterova, E. I., Andronova, T. M., & Yakubovskaya, R. I. (2017). Efficacy of combined treatment of experimental tumors with cytostatic agents and GMDP-A. *Russian Journal of Biotherapy, 16*(2), 13–22. https://doi.org/10.17650/1726-9784-2017-16-2-13-22

43. Vorontsova, M. S., Karmakova, T. A., Plotnikova, E. A., Morozova, N. B., Abakumov, M. A., Yakubovskaya, R. I., & Alexeev, B. Y. (2018). Subcutaneous and orthotopic xenograft models of human bladder carcinoma in nude mice for epidermal growth factor receptor-targeted treatment. *Russian Journal of Biotherapy, 17*(2), 31–40. https://doi.org/10.17650/1726-9784-2018-17-2-31-40

44. Machulkin, A. E., Skvortsov, D. A., Ivanenkov, Y. A., Ber, A. P., Kavalchuk, M. V., Aladinskaya, A. V., Uspenskaya, A. A., Shafikov, R. R., Plotnikova, E. A., Yakubovskaya, R. I., Nimenko, E. A., Zyk, N. U., Beloglazkina, E. K., Zyk, N. V., Koteliansky, V. E., & Majouga, A. G. (2019). Synthesis and biological evaluation of PSMA-targeting paclitaxel conjugates. *Bioorganic & Medicinal Chemistry Letters, 29*(16), 2229–22235. https://doi.org/10.1016/j.bmcl.2019.06.035

45. Fadeev, R., Chekanov, A., Solovieva, M., Bezborodova, O., Nemtsova, E., Dolgikh, N., Fadeeva, I., Senotov, A., Kobyakova, M., Evstratova, Y., Yakubovskaya, R., & Akatov, V. (2019). Improved anticancer effect of recombinant protein izTRAIL combined with sorafenib and peptide iRGD. *International Journal of Molecular Sciences, 20*(3), 525. https://doi.org/10.3390/ijms20030525

46. Petriev, V. M., Tishchenko, V. K., Smoryzanova, O. A., Mikaylovskaya, A. A., Bol'bit, N. M., Duflot, V. R., Gayvoronsky, A. V., Morozova, N. B., & Yakubovskaya, R. I. (2018). A new radiopharmaceutical,Sm-153-labelled thermoresponsive polymer, for local radiotherapy of solid tumors. *Radiation and Risk, 27*(1), 66–76. https://doi.org/10.21870/0131-3878-2018-27-1-66-76

47. Rozenkranz, A. A., Slastnikova, T. A., Karmakova, T. A., Vorontsova, M. S., Morozova, N. B., Petriev, V. M., Abrosimov, A. S., Khramtsov, Y. V., Lupanova, T. N., Ulasov, A. V., Yakubovskaya, R. I., Georgiev, G. P., & Sobolev, A. S. (2018). Antitumor activity of auger electron Emitter[111] in delivered by modular nanotransporter for treatment of bladder cancer with EGFR overexpression. *Frontiers in Pharmacology, 9*, 1331. https://doi.org/10.3389/fphar.2018.01331

48. Morozova, N. B., Plotnikova, E. A., Plyutinskaya, A. D., & Stamova, V. O. (2018). Preclinical trial of Bacteriosens used for the photodynamic therapy of malignant tumors, including prostate cancer. *Russian Journal of Biotherapy, 17*(3), 55–64. https://doi.org/10.17650/1726-9784-2018-17-3-55-64

49. Yakubovskaya, R. I., Morozova, N. B., Pankratov, A. A., Kazachkina, N. I., Plyutinskaya, A. D., Karmakova, T. A., Andreeva, T. N., Venediktova, Y. B., Plotnikova, E. A., Nemtsova, E. R., Sokolov, V. V., Filonenko, E. V., Chissov, V. I., Kogan, B. Y., Butenin, A. V., Feofanov, A. V., & Strakhovskaya, M. G. (2015). Experimental photodynamic therapy: 15 years of development. *Russian Journal of General Chemistry, 85*(1), 217–239. https://doi.org/10.1134/S1070363215010405

50. Yakubovskaya, R. I., Plotnikova, E. A., Plutinskaya, A. D., Morozova, N. B., Chissov, V. I., Makarova, E. A., Dudkin, S. V., Lukyanets, E. A., & Vorozhtsov, G. N. (2014). Photophysical properties and in vitro and in vivo photoinduced antitumor activity of cationic salts of meso-tetra(N-alkyl-3-pyridyl)bacteriochlorins. *Journal of Photochemistry and Photobiology. B, 130*, 109–114. https://doi.org/10.1016/j.jphotobiol.2013.10.017

51. Bezborodova, O. A., Alekseenko, I. V., Nemtsova, E. P., Pankratov, A., Filyukova, O., Yakubovskaya, R., Kostina, Potapov, V., & Sverdlov, E. (2018). Antitumor efficacy of a complex based on a binary system of vectors for co-expression of the killer gene FCU1 and Cre-recombinase. *Reports of the Academy of Sciences, 483*(3), 337–340. https://doi.org/10.31857/S086956520003261-0

52. Bezborodova, O. A., Nemtsova, E. R., Gevorkov, A. R., Boyko, A. V., Venediktova, J. B., Alekseenko, I. V., Kostina, M. B., Monastyrskaya, G. S., Sverdlov, E. D., Khemelevskiy, E. V., & Yakubovskaya, R. I. (2016). Antitumor efficacy of combined gene and radiotherapy in animals. *Doklady. Biochemistry and Biophysics, 470*(4), 471–474. https://doi.org/10.7868/S0869565216280227

53. Bezborodova, O. A., Nemtsova, E. P., Venediktova, Y. B., et al. Experimental gene suicidal antitumor therapy: Development of an effective treatment regimen on the model of mouse sarcoma. *Biopharmaceutical Journal, 8*(2), 40–46.

54. Karmakova, T. A., Bezborodova, O. A., Nemtsova, E. P., et al. Assessment of antitumor efficacy and mechanism of action of suicidal gene therapy in an experimental model of mouse sarcoma. *Biopharmaceutical Journal, 8*(1), 54–63.

Sustainability of Biobanks and Biobanking in LMICs

26

Daniel Simeon-Dubach and Zisis Kozlakidis

Abstract

Ensuring sustainability is the most important task of a biobank, because biobanks are very resource-intensive infrastructures with a long-term horizon. Only a sustainable biobank can make a significant contribution to research ongoing. However, many different factors must fit and interact to enable sustainability. In this chapter, the particular challenges of biobank sustainability and their biobanking activities with a special focus on LMICs will be discussed.

Keywords

Sustainability · Biobanking · Low-and middle-income countries · Challenges and opportunities · Biobanking stages

Introduction

Ensuring sustainability is the most important task of a biobank, because biobanks are very resource-intensive infrastructures with a long-term horizon. Only a sustainable biobank can make a significant contribution to research ongoing. However, many different factors must fit and interact to enable sustainability. In this chapter, we will discuss the particular challenges of biobank sustainability and their biobanking activities with a special focus on LMICs.

D. Simeon-Dubach (✉)
Medservice, Walchwil, Switzerland
e-mail: daniel.simeon-dubach@medservice.ch

Z. Kozlakidis
International Agency for Research on Cancer, World Health Organization, Lyon, France

Definition of a Biobank

Before sustainability of biobanks can be discussed, especially in LMICs settings, the characteristics of the biobank environment should be explained. A major challenge remains in that there is still no generally accepted definition of what a biobank is. This can be indicative of a rapidly growing field; however, developments in the field of biobanking will be complicated in the longer-term if researchers are unaware that their collection is indeed a biobank and that it needs to follow particular sets of best practices and international standards [1, 2]. A clear definition of the term is therefore an important step towards fostering collaboration amongst researchers, the discoverability and utilization of samples [3]. Additionally, a variation in definitions between different stakeholders and/or geographies—if continued—is likely to generate misunderstandings about the scope of various regulations and guidelines related to sample collections [4]. This is especially important in LMICs settings, where a number of regulatory frameworks have been created in recent years and/or are still in the process of being created [5, 6].

To this end, a biobank can be defined as a systematic collection of samples and data for a defined purpose. The data itself has three dimensions: (1) the data of the donor (e.g., patient and medical history; in the case of the latter, there may be various diseases that are more or less important), (2) the data of the collection process and (3) the data which is generated by using the samples. It can be stated that as soon as the samples are removed from their usual preservation environment, a new "life in data" for these samples begins. One such extreme example is immortal cell lines.

Biobanks service a wide range of stakeholders and research needs, and therefore various models of biobanks exist. From the users' point of view, there are biobanks that have only one user, few or many (mono-, oligo-, and poly-users) [7]. It is also important whether a biobank is designed for a short period of time, e.g., for the lifetime of a specific project, or for the long term. Lastly, another significant parameter is whether it is a retrospective or prospective collection [8].

Concept of "Biobanking 3.0"

The above definition of biobanks is static and does not allow for the necessary flexibility demanded by rapid technological and experimental needs. On the contrary, the concept "Biobanking 3.0" shows the dynamic components of biobanks and all the activities associated with them, so-called biobanking [9]. This concept takes into account both biobanks that are newly established and activities that are carried out when an established biobank opens up to a new field. As such, this might be a better suited concept for LMIC settings, where the pressure for flexible and inter-linking infrastructures is at its greatest. At the beginning of the COVID-19 pandemic, this could be exemplified by those biobanks that had to adapt to the new healthcare context and collect corresponding samples and data from COVID-19 patients within a very short time [10].

In the Biobanking 3.0 concept, five individual activities of biobanks are listed and include (1) the samples that are collected, (2) the personal data that are collected, (3) the sample-related data that are collected, (4) the customer need and (5) sustainability.

At the initial biobanking 1.0 stage, i.e. when a biobank is newly established or a new field is newly established within an existing biobank, the focus is on the quantity or on how many samples can be collected. At this point, it is central that contacts are established with those structures where the corresponding donors can be found and samples are available. Data are also important, but most of them can be collected with a delay. At the biobanking 2.0 stage, the focus is on the quality, and especially the quality of the data collected. Biobanking 1.0 and 2.0 can go hand in hand, and these are usually very short stages. Much more important is the stage of Biobanking 3.0 with the focus on customer needs and sustainability, where biobanks reach operational maturity. This concept gains further momentum when it is used to describe and, if necessary, calculate impact and value.

The Three Dimensions of Sustainability of Biobanks and Biobanking

The basic principles of sustainability are universally valid and applicable across different geographies and fields of activity. This well-established three-pillar concept of sustainable development has already been adapted for biobanks [9]. According to this concept, biobanks can achieve long-term sustainability if the financial, operational and social dimensions are not played off against each other, but are pursued with equal priority [11].

The *financial dimension* involves not only access to finances but also knowing the various sources of finance and how to approach them. These sources of finance can be local, regional, national or international, they can be public/governmental or private, or a mixture of these. But it is equally important to know the real costs of operating a biobank. It can be very difficult to calculate these costs because, as mentioned above, biobanks are very diverse and are integrated into a wide variety of administrative networks [12, 13]. A helpful tool is the Biobanking 3.0 concept mentioned above.

The *operational dimension* of sustainability includes all aspects of how the biobank is managed as a research infrastructure. In this context, a business plan which includes a vision and mission; an analysis of SWOT; risk mitigation; and defined performance metrics including all key business plan elements and monitoring their success against SMART (Specific, Measurable, Achievable, Relevant, Time bound) goals and objectives is a central steering tool. The business plan makes it possible to identify the opportunities that a biobank has and to take the necessary steps to exploit these opportunities [14]. All aspects of the Quality Management System need also to be included in this dimension [15].

The *social dimension* focuses on early and frequent interactions with all stakeholders. These continuous interactions increase or establish social trust and thus ultimately the general value of the biobank [16, 17]. Within LMICs settings, the social dimension has proven particularly important for collections of samples, especially as a number of past experiences were mishandled. This applies in particular in cases where samples were collected within LMICs settings, shipped abroad for the sole academic benefit of external collaborators and without involving the communities and researchers where those samples originated from [18].

Sustainability for Biobanks in LMIC: Challenges and Opportunities

Our analysis does not concern individual biobanks; as mentioned above, the variability within biobanks is enormous. Instead, this chapter focuses on the biggest challenges and opportunities for biobanking in LMICs.

Probably the biggest current challenges are infrastructure and governance [18–21] aspects influencing each other. A biobank is a long-term infrastructural commitment that requires a high level of investment, as well as a structured and continuous environment. Biobanks require buildings that should meet defined requirements, and biobanks need dedicated personnel who ensure continuous operation. The previous experiences of disconnected islands of scientific excellence across LMICs, created for the life cycle of particular research initiatives, would need to be turned into interconnected networks or collaborations [22, 23], where biobanks are the common ground of access to biological resources and data. Such networks increase the resilience of investments, can support research and capacity building and can even mitigate the risk to the overall scientific advance, if one of their participating infrastructures faces financial distress [24, 25].

Ethical, legal, social and regulatory aspects are key challenges for the successful management of biobanks. These are elements that have seen very active research in the last few years, providing evidence for informed governance structures within LMICs [5, 18, 20, 26–30]. It must also be taken into account that biomedical research and, as part of that, biobanking usually does not constitute the first governmental priority, especially as the healthcare system is not primarily oriented towards research [31, 32].

However, there are also many opportunities for biobanks in LMICs, as there are many understudied populations and many diseases with unmet need. In many countries, large cohorts of patients/donors can be accessed [33], and there is general willingness from the wider public to participate in medical research through biobanks [34–36]. Additionally, in LMICs there is a demand for sufficient, educated manpower and a growing young generation of increasingly well-educated, well-connected professionals exists. This will lead to newly established research groups who will generate a considerable demand for samples, data and other biobank services.

Outlook

Many LMICs are developing legal frameworks for research in general and specifically also for biobanking. Biobanks need to be in close contact with these legislative efforts to ensure sustainability is included or protected through such frameworks. A key success factor is good business planning for each individual biobank. While this holds true for all biobanks globally, LMICs cannot afford failing biobanks due to the significant investment needed. LMIC biobankers can learn from the experiences of others and reach for dedicated expertise beyond the timeframes of individual grants/projects. They should build local, regional or national networks and consortia to become an important stakeholder in all aspects of research and biobanking [5, 19, 33, 37, 38]. There are great opportunities for scientific collaboration by linking to existing infrastructure in LMICs and globally. This includes international organizations and societies like IARC, ISBER, or others for staff educational opportunities, e.g. qualifications and long-distance learning [39, 40].

In summary, there are good reasons to be optimistic about the future prospects of biobanking in LMIC. In addition, biobanks in LMICs can achieve long-term sustainability if the financial, operational and social dimensions are treated with equal priority, and the impact of such biobanks is documented. In this manner, successful biobanks will eventually generate demonstrable value to medical research and general health.

Disclaimer Where authors are identified as personnel of the International Agency for Research on Cancer/WHO, the authors alone are responsible for the views expressed in this article, and they do not necessarily represent the decisions, policy or views of the International Agency for Research on Cancer/WHO.

References

1. ISO 20387:18 Biotechnology – Biobanking – General Requirements for Biobanking. (2018). Accessed March 08, 2021, from http://www.iso.org/standard/67888.html
2. ISBER Best Practices. (2018). (4th ed.). Accessed March 08, 2021, from www.isber.org/bestpractices
3. Shaw, D. M., Elger, B. S., & Colledge, F. (2014). What is a biobank? Differing definitions among biobank stakeholders. *Clinical Genetics, 85*(3), 223–227. https://doi.org/10.1111/cge.12268
4. Hewitt, R., & Watson, P. H. (2013). Defining biobank. *Biopreservation and Biobanking, 11*(5), 309–315. https://doi.org/10.1089/bio.2013.0042
5. Nansumba, H., Ssewanyana, I., Tai, M., & Wassenaar, D. (2020). Role of a regulatory and governance framework in human biological materials and data sharing in National Biobanks: Case studies from biobank integrating platform, Taiwan and the National Biorepository, Uganda. *Wellcome Open Research, 1,* 4(171). https://doi.org/10.12688/wellcomeopenres.15442.2
6. Vodosin, P., Jorgensen, A. K., & Mendy, M. (2021). Review of regulatory frameworks governing biobanking in lower and middle income member countries of BCNet. *Biopres Biobank, 19*(5), 444–452.

7. Watson, P. H., & Barnes, R. O. (2011). A proposed schema for classifying human research biobanks. *Biopreservation and Biobanking, 9*(4), 327–333. https://doi.org/10.1089/bio.2011. 0020

8. Watson, P. H., Nussbeck, S. Y., Cater, C., O'Donoghue, S., Cheah, S., Matzke, L. A. M., Barnes, R. O., Bartlett, J., Carpenter, J., Grizzle, W. E., Johnston, R. N., Mes-Masson, A.-M., Murphy, L., Sexton, K., Shepherd, L., Simeon-Dubach, D., Zeps, N., & Schacter, B. (2014). A framework for biobank sustainability. *Biopreservation and Biobanking, 12*(1), 60–68. https:// doi.org/10.1089/bio.2013.0064

9. Simeon-Dubach, D., & Watson, P. (2014). Biobanking 3.0: Evidence based and customer focused biobanking. *Clinical Biochemistry, 47*(4–5), 300–308. https://doi.org/10.1016/j. clinbiochem.2013.12.018

10. Henderson, M. K., Kozlakidis, Z., Fachiroh, J., Wiafe, A. B., Xu, X., Ezzat, S., Wagner, H., Maques, M. M. C., & Yadav, B. K. (2020). The responses of biobanks to COVID-19. *Biopreservation and Biobanking, 18*(6), 483–491. https://doi.org/10.1089/bio.2020.29074.mkh

11. Simeon-Dubach, D., & Henderson, M. K. (2014). Sustainability in biobanking. *Biopreservation and Biobanking, 12*(5), 287–291. https://doi.org/10.1089/bio.2014.1251

12. Rao, A., Vaught, J., Tulskie, B., Olson, D., Odeh, H., McLean, J., & Moore, J. M. (2019). Critical financial challenges for biobanking: Report of a National Cancer Institute study. *Biopreservation and Biobanking, 17*(2), 129–138. https://doi.org/10.1089/bio.2018.0069

13. Doucet, M., Yuille, M., Georghiou, L., & Dagher, G. (2017). Biobank sustainability: Current status and future prospects. *Journal of Biorepository Science for Applied Medicine, 2017*(5), 1–7. https://doi.org/10.2147/BSAM.S100899

14. Fachiroh, J., Dwianingsih, E. K., Wahdi, A. E., Pramatasari, F. L., Hariyanto, S., Pastiwi, N., Yunus, J., Mendy, M., Scheerder, B., & Lazuardi, L. (2019). Development of a biobank from a legacy collection in Universitas Gadjah Mada, Indonesia: Proposed approach for centralized biobank development in low-resource institutions. *Biopreservation and Biobanking, 17*(5), 387–394. https://doi.org/10.1089/bio.2018.0125

15. Ferdyn, K., Glenska-Olender, J., Witon, M., Zagorska, K., Kozera, L., Chroscicka, A., & Matera-Witkiewicz, A. (2019). Quality management system in the BBMRI.Pl consortium: Status before the formation of the polish biobanking network. *Biopreservation and Biobanking, 17*(5), 401–409. https://doi.org/10.1089/bio.2018.0127

16. Jao, I., Kombe, F., Mwalukore, S., Bull, S., Parker, M., Kamuya, D., Molyneux, S., & Marsh, V. (2015). Research stakeholders' views on benefits and challenges for public health research data sharing in Kenya: The importance of trust and social relations. *PLoS One, 10*(9), e0135545. https://doi.org/10.1371/journal.pone.0135545

17. Moodley, K., & Singh, S. (2016). "It's all about trust": Reflections of researchers on the complexity and controversy surrounding biobanking in South Africa. *BMC Medical Ethics, 17*(1), 57. https://doi.org/10.1186/s12910-016-0140-2

18. Fernando, B., King, M., & Sumathipala, A. (2019). Advancing good governance in data sharing and biobanking – International aspects. *Wellcome Open Research, 4*, 184. https://doi.org/10. 12688/wellcomeopenres.15540.1

19. Mendy, M., Caboux, E., Sylla, B., Dillner, J., Chinquee, J., & Wild, C. (2014). BCNet survey participants. Infrastructure and facilities for human biobanking in low- and middle-income countries: A situation analysis. *Pathobiology, 81*, 252–260. https://doi.org/10.1159/000362093

20. Zawati, M. N., Tasse, A. M., Mendy, M., Caboux, E., Lang, M., & on Behalf of Biobank and Cohort Building Network Members. (2018). Barriers and opportunities in consent and access procedures in low-and middle-income country biobanks: Meeting notes from the BCNet training and general assembly. *Biopreservation and Biobanking, 16*(3), 171–178. https://doi. org/10.1089/bio.2017.0081

21. Bull, S., & Bhagwandins N. (2020). The ethics of data sharing and biobanking in health research. *Wellcome Open Research, 5*, 270. https://doi.org/10.12688/wellcomeopenres.16351.

22. Franzen, S. R., Chandler, C., & Lang, T. (2017). Health research capacity development in low and middle income countries: Reality or rhetoric? A systematic meta-narrative review of the qualitative literature. *BMJ Open, 7*(1), e012332. https://doi.org/10.1136/bmjopen-2016-012332

23. Ghaffar, A., Ijsselmuiden, C., & Zicker, F. (2008). *Changing mindsets. Research capacity strengthening in low- and middleincome countries.* Council on Health Research for Development (COHRED).

24. Blanchet, K., Nam, S.L., Ramalingam, B., Pozo-Martin, F. (2017). Governance and capacity to manage resilience of health systems: Towards a new conceptual framework. *International Journal of Health Policy Management, 6*(8), 431–435. https://doi.org/10.15171/ijhpm.2017.36

25. Stewart, R., El-Harakeh, A., & Cherian, S. A. (2020). Evidence synthesis communities in low-income and middle-income countries and the COVID-19 response. *Lancet, 396*(10262), 1539–1541. https://doi.org/10.1016/S0140-6736(20)32141-3

26. Merdad, L., Aldakhil, L., Gadi, R., Assidi, M., Saddick, S. Y., Abuzenadah, A., Vaught, J., Buhmeida, A., & Al-Qahtani, M. H. (2017). Assessment of knowledge about biobanking among healthcare students and their willingness to donate biospecimens. *BMC Medical Ethics, 18*(32). https://doi.org/10.1186/s12910-017-0195-8

27. He, N., Guo, Y., He, M., Qiang, W., & Li, H. (2017). Attitudes and perceptions of cancer patients toward biospecimen donation for cancer research: A cross-sectional survey among Chinese cancer patients. *Biopreservation and Biobanking, 15*(4), 366–374. https://doi.org/10.1089/bio.2016.0079

28. Tindana, P., & de Vries, J. (2016). Broad consent for genomic research and biobanking: Perspectives from low- and middle-income countries. *Annual Review of Genomics and Human Genetics, 17*, 375–393. https://doi.org/10.1146/annurev-genom-083115-022456

29. Tindana, P., Yakubu, A., Staunton, C., Matimba, A., Littler, K., Madden, E., Munung, N. S., de Vries, J., & Members of the H3Africa Consortium. (2019). Engaging research ethics committees to develop an ethics and governance framework for best practices in genomic research and biobanking in Africa: The H3Africa model. *BMC Medical Ethics, 20*(1), 69. https://doi.org/10.1186/s12910-019-0398-2

30. Gao, H., Jiang, J., Feng, B., Guo, A., Hong, H., & Liu, S. (2018). Parental attitudes and willingness to donate children's biospecimens for congenital heart disease research: A cross-sectional study in Shanghai, China. *BMJ Open, 8*(10), e022290. https://doi.org/10.1136/bmjopen-2018-022290

31. Goodyear-Smith, F., Bazemore, A., Coffman, M., Fortier, R. D. W., Howe, A., Kidd, M., Phillips, R., Rouleau, K., & van Weel, C. (2019). Research gaps in the organisation of primary healthcare in low-income and middle-income countries and ways to address them: A mixed-methods approach. *BMJ Global Health, 4*(Suppl 8), e001482. https://doi.org/10.1136/bmjgh-2019-001482

32. Goodyear-Smith, F., Bazemore, A., Coffman, M., Fortier, R., Howe, A., Kidd, M., Phillips, R., Rouleau, K., & van Weel, C. (2019). Primary care financing: A systematic assessment of research priorities in low- and middle-income countries. *BMJ Global Health, 4*(Suppl 8), e001483. https://doi.org/10.1136/bmjgh-2019-001483

33. Akinyemi, R. O., Akinwande, K., Diala, S., Adeleye, O., Ajose, A., Issa, K., Owusu, D., Boamah, I., Yahaya, I. S., Jomoh, A. O., Imoh, L., Fakunle, G., Akpalu, A., Sarfo, F., Wahab, K., Sanya, E., Owolabi, L., Obiako, R., Osaigbovo, G., et al. (2018). Biobanking in a challenging African environment: Unique experience from the SIREN project. *Biopreservation and Biobanking, 16*(3), 217–232. https://doi.org/10.1089/bio.2017.0113

34. Abdelhafiz, A. S., Sultan, E. A., Ziady, H. H., Ahmed, E. O., Khairy, W. A., Sayed, D. M., Zaki, R., Fouda, M. A., & Labib, R. M. (2019). What Egyptians think. Knowledge, attitude, and opinions of Egyptian patients towards biobanking issues. *BMC Medical Ethics, 20*(57). https://doi.org/10.1186/s12910-019-0394-6

35. Mezinska, S., Kaleja, J., Mileiko, I., Santare, D., Rovite, V., & Tzivian, L. (2020). Public awareness of and attitudes towards research biobanks in Latvia. *BMC Medical Ethics, 21*(1), 65. https://doi.org/10.1186/s12910-020-00506-1

36. Lhousni, S., Daoudi, F., Belmokhtar, I., Belmokhtar, K. Y., Abda, N., Boulouiz, R., Tajir, M., Bellaoui, M., & Quarzane, M. (2020). Patients' knowledge and attitude toward biobanks in Eastern Morocco. *Biopreservation and Biobanking, 18*(3), 189–195.
37. De Oliveira, L., Dias, M. A. B., Jeyabalan, A., Payne, B., Redman, C. W., Magee, L., Poston, L., Chappell, L., Seed, P., von Dadelszen, P., & Roberts, J. M. (2018). Creating biobanks in low and middle-income countries to improve knowledge – The PREPARE initiative. *Pregnancy Hypertens, 13*, 62–44. https://doi.org/10.1016/j.preghy.2018.05.007
38. Sumathipala, A., Siribaddana, S., Hotopf, M., McGuffin, P., Glozier, N., Ball, H., et al. (2013). The Sri Lankan twin registry: 2012 update. *Twin Research and Human Genetics, 16*(1), 307–312. https://doi.org/10.1017/thg.2012.119
39. Henderson, M. K., & Kozlakidis, Z. (2018). ISBER and the biobanking and cohort network (BCNet): A strengthened partnership. *Biopreservation and Biobanking, 16*(5), 393–394. https://doi.org/10.1089/bio.2018.29043.mkh
40. Vaught, J. (2016). Biobanking and biosecurity initiatives in Africa. *Biopreservation and Biobanking, 14*(5), 355–356. https://doi.org/10.1089/bio.2016.29009.jjv

Biobanks for Enabling Research and Development by Trusted Patient Data Environment

27

Bernhard Zatloukal, Heimo Müller, Werner Strasser, and Kurt Zatloukal

Abstract

Digital transformation of health systems will critically rely on accessing data from patients in order to develop technologies and for their future implementation in health service. In this context, it is essential that data from relevant patient groups have been used to develop technologies since otherwise the performance and patient safety of digital technologies cannot be guaranteed. For this reason, a trusted environment for health data has to be established comprising several elements. It has to build on (1) a legal basis (e.g., the European General Data Protection Regulation), (2) be centered on patient engagement, (3) employ innovative privacy preserving technologies, (4) apply proper governance models, (5) ensure data sovereignty, and (6) allow fair benefit sharing from using patient data. Furthermore, new data access models should be explored that enable research on health data in the trusted environment and avoid the need of sending data outside this environment to users.

Keywords

Digital transformation · Health systems · Trusted data environment · Cloud services · Secret sharing

B. Zatloukal · W. Strasser
fragmentiX Storage Solutions GmbH, Klosterneuburg, Austria

H. Müller · K. Zatloukal (✉)
Diagnostic and Research Center for Molecular BioMedicine, Diagnostic and Research Institute of Pathology, Medical University of Graz, Graz, Austria
e-mail: kurt.zatloukal@medunigraz.at

Introduction

Health care increasingly relies on the proper use of a broad spectrum of data requiring innovative solutions for efficient and secure data management and analysis. In this context, machine learning and AI provide new opportunities to increase capacities and competencies of health care workers which are essential to sustain health systems. In several medical fields, there is a global shortage of trained medical specialists, which particularly affects developing countries. For example, there are countries with less than one pathologist per nine million people, which prevents patients with cancer from getting a proper diagnosis, and without proper diagnosis there will be no proper therapy even when medication would be available. There is increasing evidence that AI can assist medical professionals in diagnostics and increases their working capacity, which provides new opportunities to cope with global shortage of trained medical specialists.

This importance of using health data for future health systems is highlighted in the report of the World Health Organization in its Global Strategy on Digital Health 2020–2025 by the statement "Digital transformation of health care can be disruptive; however, technologies such as the Internet of things, virtual care, remote monitoring, artificial intelligence, big data analytics, blockchain, smart wearables, platforms, tools enabling data exchange and storage and tools enabling remote data capture and the exchange of data and sharing of relevant information across the health ecosystem creating a continuum of care have proven potential to enhance health outcomes by improving medical diagnosis, data-based treatment decisions, digital therapeutics, clinical trials, self-management of care and person-centred care as well as creating more evidence-based knowledge, skills and competence for professionals to support health care" [1]. A specific challenge in the potentially disruptive innovation in health systems is that low-income countries are not left behind, which may create major inequalities for people for accessing high-quality and affordable health care as defined in the Sustainable Development Goal 3 of the United Nations (UN SDG3) [2].

The advancement of digital health, particularly the development tools involving machine learning algorithms and AI, critically relies on accessing health-related data from a broad spectrum of people and disease conditions. The importance of inclusiveness and quality of health data for proper performance of digital health technologies is underlined in the report of the UN Secretary-General's High-level Panel on Digital Cooperation emphasizing "Gaps in the data on which algorithms are trained can likewise automate existing patterns of discrimination, as machine learning systems are only as good as the data that is fed to them" [3]. This need of making health data available from patients around the globe goes hand in hand with the need of creating a trusted environment where these data can be handled properly as further stated by UN Secretary-General's High-level Panel on Digital Cooperation emphasizing "We recommend the development of a Global Commitment on Digital Trust and Security to shape a shared vision, identify attributes of digital stability, elucidate and strengthen the implementation of norms for responsible uses of technology, and propose priorities for action" [3].

Proposed Solution

In order to build such trusted environment for health data, biobanks could play a central role. In context of establishing biobank research infrastructures, which provide access to quality controlled human biosamples and associated medical data for research in compliance with ethical and legal requirements, several aspects for such trusted environment have been developed [4–8]. Key lessons learned are that a trusted environment requires a multi-pillar approach that builds on the applicable legal framework (e.g., the GDPR) and has to be centered on the needs and interests of patients. Furthermore, it should include several elements such as privacy preserving technology, governance models ensuring data sovereignty, and fair benefit sharing for providing access to data (Fig. 27.1). The latter is of particular importance for low-income countries because in these countries, data analysis capacities are often not sufficiently available so that the value of analyzing data (i.e., knowledge generation and product development) is generated outside of the country that provided the data.

It is very important that health data coming either from hospitals or directly from patients (e.g., provided by health apps) are stored and processed in a secure environment in order to ensure privacy of patients and to make data available for research and innovation. In this context cloud solutions gain increasing importance.

The low availability of suitably and sufficiently qualified IT staff worldwide makes it very difficult, especially in regions with low wage levels, to build up the infrastructure to support modern medical applications within their own digital sphere of sovereignty. The use of cloud-based services to perform computing tasks as well as to store data in cloud-based storage systems therefore often seems to be without alternatives. Too often, the personal data of a large group of people—up to the entire gene pool of entire nations—is transferred to a single cloud provider for processing and storage. By transferring this data in ultimately readable form, there is a very substantial risk that this data could be leaked and used against the interests of the patients. To consistently maintain true digital sovereignty over such sensitive data,

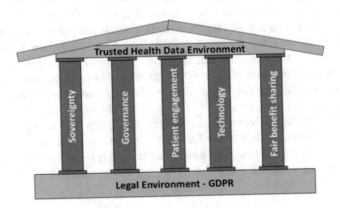

Fig. 27.1 Key features of a trusted health data environment

contractual assurances and bilateral trust are not sufficient. With the help of scientifically substantiated methods and their technologically optimized use, a cryptographic guarantee must be made possible that can also withstand the soon to be more widely available quantum computers. Particularly in less sophisticated healthcare systems, the use of external resources is often unavoidable. Even if storage space or CPU time is often provided free of charge to fight diseases, it must be assumed that medical datasets may also be forwarded to others or copied for undisclosed purposes. However, by using suitable technologies, it is possible to use cloud-based resources in such a way that patients' data do not become the currency to pay for "free" resources.

The well-established cryptographic algorithm known as Shamir's Secret Sharing (SSS) may be able to provide the above mentioned requirements for securely using public cloud storage for sensitive medical data. SSS is a (k, n)-threshold scheme which splits data into multiple fragments: n data fragments are generated and k fragments are required to reconstruct the original data. SSS can provide information-theoretic security (ITS) for data fragments at rest (e.g., on cloud storage) as well as resilience against data loss (e.g., infrastructure failure or cyber threats). If individual data fragments are lost or stolen or otherwise damaged by an attacker, a full reconstruction from the remaining k data fragments is possible. At the same time, knowledge of less than k fragments leaves the original data completely undetermined, i.e., it is impossible to gain any knowledge about the original data from any number of fragments lower than the minimum needed for reconstruction (k fragments). The encryption cannot be broken even with future (quantum) computers and infinitely high computing capacity. It is mathematically "perfectly" secure and is, therefore, called Perfect Secret Sharing (PSS) [9]. These security features are now made available to users as complete plug-and-play solution (e.g., fragmantiX storage appliances) to be easily integrated into an existing network infrastructure and offer additional features achieved by combining different encryption approaches (Fig. 27.2). Operators can then choose the (k, n)-threshold as well as where the data fragments are stored. Furthermore, they can use different storage providers and upload each fragment via a different internet service provider. As a result, the operator does not have to trust a single storage provider or internet provider, but can distribute the fragments internationally and can use cheaper storage providers from different countries without concern. The higher total storage requirements for the sum of all data fragments compared to just the original data can be reduced by choosing cheaper storage providers, because one does not have to rely on the integrity of a single provider. In addition, storage requirements can be further reduced by combining SSS with other cryptographic algorithms. This is called Computationally Secure Secret Sharing (CSS) [10, 11]. Compared to PSS, CSS does not provide ITS, but it is also reasonably secure against future (quantum) computers. It is possible to choose which data is fragmented with PSS (e.g., highly sensitive genetic data) and which data is fragmented with CSS (e.g., large medical

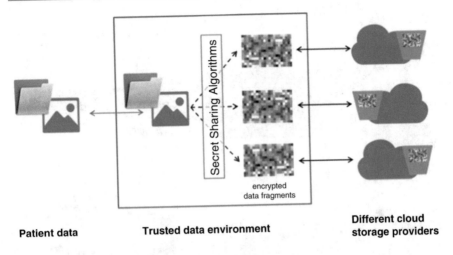

Patient data   **Trusted data environment**   **Different cloud storage providers**

Fig. 27.2 Principle of secret sharing for cloud solutions

image data with high storage requirements) and thus individually create different storage strategies for different scenarios.

Another aspect to be considered is that currently access to health data is provided to researchers by sending data to the requestor after the data access application has been evaluated for scientific relevance as well as ethical and legal compliance. However by sending the data to the requestor data leave the trusted environment, which may undermine the trust given by patients in context of consenting and donation of data for research. In order to avoid this issue, solutions should be implemented that keep the data in the secure and trusted homeland environment for analysis which means that data do not need to be sent abroad to researcher but the researcher analyses the data in the original environment (bringing the algorithms to the data instead of sending the data to the algorithms) (Fig. 27.3).

The rapid increase in the computing power of modern computer systems and the simultaneous optimization of the algorithms starting in the late 1970s [12] and data handling paradigms used in medicine, among other domains, have made it possible to use Secure Multi-party Computation (SMPC) methods in more and more areas of application. This enables the protection of, on the one hand, personal medical data and, on the other hand, the protection of the IP of the algorithm creators. By combining (1) the storage of sensitive medical datasets using secret sharing—and the resulting ITS [13]—and (2) the possibility to let others compute on this data without the need to send the data to the user, new technical solutions became available to build a trusted data environment. Even organizations, agencies or countries that would otherwise be in competition with each other will be able to use these methods to cooperate on specific issues.

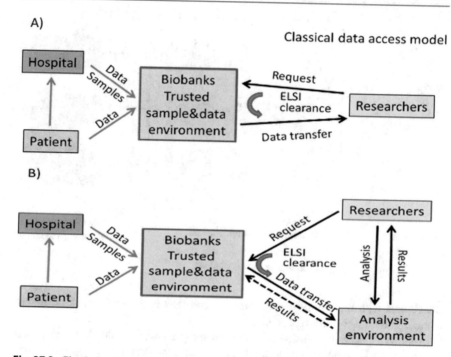

Fig. 27.3 Classical and new data access models. (**a**) Classical data access model where data is sent to the researcher. (**b**) New data access model foresees that data do not leave the trusted environment and allows researchers to analyze but not to retrieve the original data

References

1. World health Organization. Global strategy on digital health 2020–2025. gs4dhdaa2a9f352b0445bafbc79ca799dce4d.pdf (who.int).
2. UN SGD3 Target 3.8 achieve universal health coverage (UHC), including financial risk protection, access to quality essential health care services, and access to safe, effective, quality, and affordable essential medicines and vaccines for all – Indicators and a Monitoring Framework, https://indicators.report/targets/3-8/.
3. UN Secretary-General's High-level Panel on Digital Cooperation, The age of digital interdependence Report, DigitalCooperation-report-for web.pdf (un.org)
4. Müller, H., Dagher, G., Loibner, M., Stumptner, C., Kungl, P., & Zatloukal, K. (2020) Biobanks for life sciences and personalized medicine: Importance of standardization, biosafety, biosecurity, and data management. *Current Opinion in Biotechnology, 65,* 45–51. https://doi.org/10.1016/j.copbio.2019.12.004.
5. Holub, P., Kohlmayer, F., Prasser, F., Mayrhofer, M. T., Schlünder, I., Martin, G. M., Casati, S., Koumakis, L., Wutte, A., Kozera, Ł., Strapagiel, D., Anton, G., Zanetti, G., Sezerman, O. U., Mendy, M., Valík, D., Lavitrano, M., Dagher, G., Zatloukal, K., van Ommen, G. B., & Litton, J. E. (2018). Enhancing reuse of data and biological material in medical research: From FAIR to FAIR-health. *Biopreserv Biobank,16* (2), 97–105. https://doi.org/10.1089/bio.2017.0110
6. Harris, J. R., Burton, P., Knoppers, B. M., Lindpaintner, K., Bledsoe, M., Brookes, A. J., Budin-Ljøsne, I., Chisholm, R., Cox, D., Deschênes, M., Fortier, I., Hainaut, P., Hewitt, R., Kaye, J.,

Litton, J. E., Metspalu, A., Ollier, B., Palmer, L. J., Palotie, A., ... Zatloukal, K. (2012). Toward a roadmap in global biobanking for health. *European Journal of Human Genetics, 20*(11), 1105–1111. https://doi.org/10.1038/ejhg.2012.96

7. Eder, J., Gottweis, H., & Zatloukal, K. (2012). IT solutions for privacy protection in biobanking. *Public Health Genomics, 15*(5), 254–262. https://doi.org/10.1159/000336663

8. Yuille, M., van Ommen, G. J., Bréchot, C., Cambon-Thomsen, A., Dagher, G., Landegren, U., Litton, J. E., Pasterk, M., Peltonen, L., Taussig, M., Wichmann, H. E., & Zatloukal, K. (2008). Biobanking for Europe. *Brief Bioinform, 9*(1), 14–24. https://doi.org/10.1093/bib/bbm050

9. Shamir, A. (1979). How to share a secret. *Communications of the ACM, 22*(11), 612–613. https://doi.org/10.1145/359168.359176

10. Rabin, M. O. (1989). Efficient dispersal of information for security, load balancing, and fault tolerance. *Journal of the ACM (JACM), 36*(2), 335–348.

11. Krawczyk, H. (1993). Secret sharing made short. In Annual International Cryptology Conference (pp. 136–146). Springer.

12. Shamir, A., Rivest, R. L., & Adleman, L. M. (1979). "Mental Poker", technical report LCS/TR-125. Massachusetts Institute of Technology.

13. Shannon, C. E. (1949). Communication theory of secrecy systems. *Bell System Technical Journal, 28*(4), 656–715. https://doi.org/10.1002/j.1538-7305.1949.tb00928.x

Future Perspective of the Biobanking Field   **28**

Christine Mitchell and Karine Sargsyan

Abstract

Setting up a biobank may seem challenging, but with the necessary scientific enthusiasm and even limited resources, it is certainly achievable. However, some important factors need to be taken into account to ensure the success and sustainability of the biobank. As already mentioned in several chapters before, the key to the success of biobanks lies in national and international cooperation and in the exchange of information and in the implementation of new skills or documents. Cooperating with established biobanks in biobank networks and transferring knowledge and know-how to young biobanks in development, including low- and middle-income countries, are also key aspects for future harmonization.

Keywords

Biobank networks · Education and training · Knowledge transfer · Young biobanks · Future harmonization and standardization

Establishing a biobank can seem challenging, but with the necessary scientific enthusiasm and even with limited resources, it is certainly achievable. However, to

C. Mitchell
International Biobanking and Education, Medical University of Graz, Graz, Austria

K. Sargsyan (✉)
International Biobanking and Education, Medical University of Graz, Graz, Austria

Department of Medical Genetics, Yerevan State Medical University, Yerevan, Armenia

Ministry of Health of the Republic of Armenia, Yerevan, Armenia
e-mail: karine.sargsyan@medunigraz.at

209

ensure the success and sustainability of a biobank, some important factors need to be taken into account.

With the growing use of standardized biological samples for biomedical research and the development of new emerging technologies and approaches, the scientists are increasingly interested in the accessibility and quality of biological samples and their associated data, as well as collection strategies. Preanalytical processes are becoming increasingly important and require additional resources. By providing a variety of different biological materials together with the associated data, biobanks play a crucial role in research for the prevention, diagnosis and therapy of diseases. Their aim is to improve the national health care and biomedical research.

The key to the success of biobanks lies in national and international cooperation, in the exchange of information and in the implementation of new skills or documents. Cooperating with established biobanks in biobank networks such as BBMRI, ISBER and ESBB and transferring knowledge and know-how to young biobanks in development, including low- and middle-income countries, are also key aspects for future harmonization [1]. Referring to Macheiner et al., it is crucial to extend the efficiency of an individual biobank to biobank networks. Harmonization steps in the processing, storage and retrieval of samples and data are essential [2].

According to Kinkorová and Topolčan, the material stored in biobanks is a treasure for future technologies that will be able to utilize the currently uncovered information and knowledge [1]. A great challenge is to increase harmonizing networking activities among various institutions, as research-intensive organizations increasingly tend to work in small teams, which creates a crowding out effect for a huge number of highly potential research organizations. Therefore, it is even more important to establish and maintain multidisciplinary scientific cooperation with experts and with other partners, both at national and international level. To achieve and maintain a secure foothold in the world of research, participation in national and international research programs is of crucial importance, as is the publication of research results in prestigious international journals recognized and appreciated by experts and to promote partnerships with other regional clinics and scientific organizations by encouraging co-authors from other regional scientific establishments to participate in planned scientific medical journals and research publications.

Biobanking is more than just a service for biomedical research, as it is an ongoing development process within an interdisciplinary field. Therefore, the sector needs training and education in biobanking for the development of new biobanks in order to promote a harmonization within biobanks as well as activities in biobanking research for the development of novel approaches for collection, storage and the utilization of biospecimens.

According to Kirsten and Hummel, biobanks have undergone a rapid development in recent years, from simple sample collections within clinics and research institutes to highly professionalized and standardized operations [3]. This is also reflected in the establishment or continuation of international associations such as BBMRI, ISBER and ESBB. Most of these organizations see their main task in harmonizing the collection, preparation and storage of biomaterials and creating standards to improve the quality of the samples themselves and to make them traceable but also to make samples of different origins comparable and exchangeable for the creation of large (international) cohorts. In addition, BBMRI-ERIC is the first infrastructure project funded by the European Union to pursue the goal of creating a pan-European biobank infrastructure through the networking of biobanks, which will make it possible in the future to provide academic and non-academic research projects with access to many biobanks via a uniform path [3].

Biobankers must consider that a biobank is only truly successful and beneficial if the stored high-quality samples are used not only for collection but also for research purposes and for the development of advanced precision medicine, diagnosis, and treatments. A balance must be maintained between the intake and dispensing of samples for a biobank to operate sustainably and effectively.

While many biobanks are financed through a variety of projects, donations and financial grants from the government, losses can arise rapidly which can lead to considerable constraints and challenges, especially in low- and middle-income countries, as government financing and grants can be halted abruptly. Securing financial security for the biobanks' everyday operations is also made more challenging by the fact that public funds often only cover the expenses of one period. The funding or financing of a biobank often poses a major challenge. On the one hand, there are high fixed costs for infrastructure and human resources to cover, and on the other hand, the innovation and spectrum extension of novel biospecimen types and the increasing number of possible applications need additional financial resources to keep pace with the future biobanking progress. Thus, a backup plan is indispensable to ensure the sustainability, maintenance and preservation of the institution, including a solid international positioning. Consequently, the continuation of the biobank must be guaranteed independent of external factors and subsidies.

Biobanks are an extraordinary tool that, over the decades, have gradually created new research platforms and new opportunities to learn more about the operation of life systems.

In the "new biological" era, the future of human biobanks should be secured for biobanking of donor samples with acute and/or chronic physical-pathological diseases, leading to promising opportunities to understand more complex processes in the human body [4]. Furthermore, it is important and promising for biodiversity system studies and state-of-the-art biological—pathological systemic investigations of human health [5]. New research capabilities are determining the duration of research and diagnosis, and biobanks are improving the time frame and quality of biomedical research, particularly through -omics technologies [6]. An infinite number of biological samples have already been collected. In the future, the associated data will be the decisive factor, both about the donor and about the bio specimen itself. The comparison and/or combination of samples and data stored in biobanks in different regions will become equally crucial. The "biomedical research world" must seize these opportunities to meet the new demands of technology. The use of new, more efficient research technologies indeed justifies the importance of high-quality biobanks and the rapid use of biospecimens and associated data for the benefit of improved human population health. Most biobanks are developed for a specific research question and designed specifically for that purpose.

The main questions regarding the future development of biobanks are related to a possible harmonization so that the development can be resolved in different ways depending on whether the biobanking community will be able to take reasonable steps towards harmonization or not. The use of multiple human biological samples (blood, saliva, biological tissue, fluids and nucleic acids) and medical data for the application of new technologies requires a novel and larger skill set. Knowledge in the areas of biology, medicine, biochemistry, biotechnology, bioinformatics, epidemiology, engineering, strategic development, ethics and finally finances is required for ensuring the high performance and profitability (not in money but in knowledge) of these large infrastructures. Biobanks encompass the collection, storage, use and dissemination of biological samples for research assuring the effective, efficient and modern research infrastructure system in medicine.

The known collections of biological samples initially originated as collections of residues from biological samples taken for diagnostic (or) medical care purposes and were then used for research purposes [7]. Over the years, biobanks evolved into designated collections, initially episodic, in proactive research laboratories, but then fully differentiated in purpose and operation. Modern biobanking is trending in the direction of scientists accepting the need for dedicated, coordinated actions for better organized collections and systems to facilitate genomics and other -omics research. This trend has prompted the integration of biomedical research infrastructures not only in private institutions. The harmonisation and integration of biobanks and the general governance of research infrastructure became a necessity [8]. Advances in existing technologies are required to improve research infrastructures such as biobanks and biorepositories not only in terms of quantity but also in quality, in ethical clearness and in operational speed. Nevertheless, in many cases, the implementation of large-scale biomedical investigations requires several comparable biobanks that are not only operationally comparable but also comparable in terms of their strategy and business type. The future patterns show the need to meet new challenges of high-quality, standardization, ethically proven data banking, working in a public–private partnership environment, high-quality standards as well as with different types of organizations. Innovative advances in biobanking are urgently needed, especially for higher effectiveness, quality assurance and improvement, as well as easier workflow implementations. The processes are becoming increasingly sophisticated, targeted, standardized and secured, especially the collection and storage methods. These methods influence the quality of samples, the amount of available material and corresponding well-annotated data, the possibility and security of identification, etc. Therefore, efforts to harmonize, especially the technical aspects of the collection, are becoming more important every day in order to create useful and effective biobanks [9]. In the past, mainly blood, urine and (frozen) tissue were used as biobank contents. Recently, however, it has become increasingly important to store new types of human biological samples with high information potential, such as circulating tumour cells, liquid biopsy of freely circulating DNA and so on. On the other hand, samples collected with less or non-invasive methods are gaining more attention, like saliva, urine, hair, stool, different microbiome, etc. It is obvious that these new materials as well as the traditional material still must be maintained with high sampling and storage standards as well as quality, and the most obvious way will be the common or comparable protocols in biobanking. Therefore, it is necessary to use the standard technical principles. The first choice would be the certification of biobanks by the ISO biobanking norm, but if it is not possible (in developing countries the reason can be financial, organisational, availability issues, etc.), there are still several standards and recommendations mentioned in this book, which can help to maintain sustainable quality.

While establishing a biobank is currently a very productive measure for biomedical research, especially in low- and middle-income countries, it is also feasible, as mentioned earlier, especially when considering the additional measures that can provide a necessary backup in case of unforeseen circumstances and events.

References

1. Kinkorová, J., & Topolčan, O. (2018). Biobanks in horizon 2020: Sustainability and attractive perspectives. *EPMA Journal, 9*, 345–353. https://doi.org/10.1007/s13167-018-0153-7
2. Macheiner, T., Huppertz, B., Bayer, M., & Sargsyan, K. (2017). Challenges and driving forces for business plans in biobanking. *Biopreservation and Biobanking ,15*(2), 121–125. https://doi.org/10.1089/bio.2017.0018
3. Kirsten, R., & Hummerl, M. (2016). Securing the sustainability of biobanks. *Bundesgesundheitsblatt – Gesundheitsforschung – Gesundheitsschutz, 59*, 390–395. https://doi.org/10.1007/s00103-015-2302-7
4. Hartman, V., Matzke, L., & Watson, P. H. (2019). Biospecimen complexity and the evolution of biobanks. *Biopreservation and Biobanking, 17*(3), 264–270. https://doi.org/10.1089/bio.2018.0120
5. Somiari, S. B., & Somiari, R. I. (2015). The future of biobanking: A conceptual look at how biobanks can respond to the growing human biospecimen needs of researchers. *Advances in Experimental Medicine and Biology, 864*, 11–27. https://doi.org/10.1007/978-3-319-20579-3_2
6. Mendy, M., Lawlor, R. T., van Kappel, A. L., Riegman, P., Betsou, F., Cohen, O. D., & Henderson, M. K. (2018). Biospecimens and biobanking in global health. *Clinics in Laboratory Medicine, 38*(1), 183–207. https://doi.org/10.1016/j.cll.2017.10.015
7. Caenazzo, L., Tozzo, P., & Pegoraro, R. (2013). Biobanking research on oncological residual material: A framework between the rights of the individual and the interest of society. *BMC Medical Ethics, 14*, 17. https://doi.org/10.1186/1472-6939-14-17
8. Caenazzo, L., Tozzo, P., & Borovecki, A. (2015). Ethical governance in biobanks linked to electronic health records. *European Review for Medical and Pharmacological Sciences, 19*(21), 4182–4186.
9. Muller, R., Betsou, F., Barnes, M. G., Harding, K., Bonnet, J., Kofanova, O., Crowe, J. H., & International Society for Biological and Environmental Repositories (ISBER) Biospecimen Science Working Group. (2016). Preservation of biospecimens at ambient temperature: Special focus on nucleic acids and opportunities for the biobanking community. *Biopreservation and Biobanking, 14*(2), 89–98. https://doi.org/10.1089/bio.2015.0022

Annexes

Annex 1: Questions of BEMT Survey

Section	Questions
Biobank Demographics	Name
	Title
	Institution Name
	Institution Type
	Biobank Name
	How many years has your biobank been in operation?
	In which country is your biobank headquarters located?
	In what geographic region is your headquarters located?
	Are your biobank operations international?
	What is your biobank's operational model?
	How many operating sites does your biobank have?
	What is the estimated total size of your biobank facility?
Cost Recovery and Funding	Does your biobank practice cost recovery?
	Approximately what percentage (%) of your biobank operations is recovered from provision of specimens, products, and/or service?
	What is the percentage contribution of each type of fund listed below to the overall funds brought in annually to the biobank?
	Of fees recovered annually by the biobank, what percentage is contributed by each of the entities below?

(continued)

K. Sargsyan et al. (eds.), *Biobanks in Low- and Middle-Income Countries: Relevance, Setup and Management*, https://doi.org/10.1007/978-3-030-87637-1

Section	Questions
	Approximately how much (in USD) (in total) did your biobank spend on "big-ticket" item expenditures at start-up?
	Approximately how much (in USD) (for each) did your biobank spend on "big-ticket" item expenditures at start-up?
	Approximately, how much (in USD) is your total annual operational budget for operational expenditures?
	Approximately how much (in USD) does your biobank spend annually on the following operational expenditures?
	Approximately what percentage of your time do you allocate annually to performing the following activities for your biobank?
Cost and Pricing	Estimated annualized salary range, fringe benefit percentage (%), and number of FTEs for employees your biobank supports
	What type of capital equipment do you have to support specimen collection and processing?
	What type of capital equipment do you have to support sample management and storage?
	What type of capital equipment do you have to support specimen quality control and sample analysis?
	How many do you have of each piece of equipment?
	Is equipment purchased or leased?
	Estimated average annual lease/purchase price of equipment
	How many years do you plan to keep the equipment bought or leased?—Estimated annual service contract price (if applicable) per unit
Specimen, Products, and Services	Details about specimens and products biobank offers to customers
	Types of services and pricing
	Approximate % mark-up for specimens/products and services
	Approximate % mark-up for overhead to external parties (who use and collaborate with biobank)
Quality Management	What best practice guidelines does your biobank practice?

Annex 2: Patient Consent Form of the UAB

Patient Consent Form

of «___»_____ 2018

to the Protocol Agreement №_____

of «___»_____ 2018

Patient Consent Form

Biological samples collection

Full name of patient:_____

Patient ID:

_____

We would like to offer you the opportunity to provide samples of your blood, urine, saliva, surplus tissue (surgically removed) (and) or surplus aspirated fluids (ascitic fluid, pleural effusion cyst aspiration, synovial fluid, and (or) bone marrow aspiration) that will be taken from you during the course of your treatment, for the purpose of research to fight cancer.

If you wish to provide such samples, please read carefully the information provided in this document. You may ask any questions that you have regarding this study.

Форма Згоди пацієнта

від «___» _____ 2018 року

до Протокольного Договору №_____

від «___» _____ 2018 року

Форма Згоди Пацієнта

Забір біологічних зразків

ПІБ
пацієнта:_____

_____

Код   пацієнта:

_____

Ми пропонуємо Вам передати частину зразків Вашої крові, сечі, слини, надлишкової тканини (видаленої під час операції), та (або) надлишкові аспіраційні рідини (асцитна рідина, плевральний випот, синовіальна рідина, аспірат кісткового мозку) які будуть взяті підчас Вашого лікування, для проведення досліджень по боротьбі з раком.

Якщо Ви виявите бажання надати такі зразки, тоді уважно прочитайте інформацію, надану Вам
у цьому документі. Ви можете вільно поставити будь-які

Study objective:

The objective of this study is to seek new methods for the diagnosis and treatment of oncology diseases. You are invited to participate in this study because your surplus tissue, blood, urine, saliva (and) or aspirated fluids may contain biomarkers of your disease and may be useful for the above-mentioned research.

You will be asked to donate blood (not more than 50 ml) (and) or urine (not more than 50 ml) which will be drawn/collected during regular testing procedures planned in the hospital where you are treated.

After surgery in the pathology laboratory of the hospital where you are treated, the study of the affected tissue will be performed. Surplus (left-over) tissue removed during the operation that is not needed for making the diagnosis may be used for this study. That is why there is no need to perform any additional surgery or manipulation to collect tissue.

Samples of your surplus tissue (and) or blood, urine, saliva, aspirated fluids, as well as your medical data, will be used for the study. They may be sent to research institutes, centers, research companies, including commercial, for-profit companies, for example, laboratories that perform samples

Мета дослідження:

Метою цього дослідження є пошук нових методів виявлення і лікування онкологічних захворювань. Вас запрошено взяти участь в цьому Дослідженні тому, що Ваша надлишкова тканина, кров, сеча, слиначи інші аспірацій ні рідини можуть містити маркери Вашого захворювання, і будуть корисними для вищезгаданого Дослідження.

Забір зразків для Дослідження проведуть одночасно із звичайним забором крові (не більше ніж 50 ml) чи забором сечі (не більше ніж 50 ml) інших типів зразків для аналізів, які заплановані у лікарні, де Ви проходите лікування.

Після операції в патологоанатомічній лабораторії тієї лікарні, де Вас лікують, буде проведено дослідження ураженої тканини. Залишки тканини, видаленої підчас операції, які не знадобились для встановлення діагнозу, можуть бути використані для даного дослідження. Тому немає жодної необхідності проводити додаткову операцію чи маніпуляції для забору тканин.

У Дослідженні будуть використані зразки Вашої надлишкової тканини, крові, сечі, слини, аспірацій ні рідини а також медична інформація стосовно Вас. Вони в подальшому можуть бути передані в дослідницькі інститути, центри, наукові, в тому числі

testing. The samples and the data may be used immediately. It is also possible that they may be kept over an unlimited period of time for future studies.

Researchers will use biological samples primarily to determine the presence of biomarkers (proteins, RNA, DNA). Mutations in genes may appear to cause the disease or the presence of some genes can also increase the risk of developing some diseases. Therefore patients' clinical data and information about their genes can promote the development of new methods of diagnosis and treatment of oncological diseases.

Procedures for sample collection:

Surplus surgical tissue: No additional surgical procedures are required to obtain tissue samples since only left-over tissue will be used for this study.

Blood: You may be asked to donate blood for this study. Blood samples may or may not be collected with other required blood draws necessary for your treatment. Depending on your health status, the amount of blood collected for this study will not exceed 50 mL during a single blood draw. You might be requested to donate blood on multiple occasions. The amount and

комерційні, компанії, наприклад, лабораторії, які проводять тестування зразків. Зразки та інформація можуть бути використані відразу. Можли во також, що їх зберігатимуть протягом необмеженого періоду часу для майбутніх досліджень.

Дослідники будуть використовувати біологічні зразки в першу чергу для виявлення біологічних маркерів захворювання (білків, РНК, ДНК). Мутації в генах можуть виявитись причиною захворювання, наявність певних генів також може становити ризик розвитку певних хвороб.
Таким чином, інформація щодо пацієнтів і їх генів може сприяти розвитку нових методів діагностики і лікування онкологічних захворювань.

Використання біологічних зразків

Надлишкова післяопераційна тканина: для дослідження буде використана надлишкова післяопераційна тканина, без додаткового хірургічного втручання.

Зразки крові: ви можете погодитись надати кров
в рамках даного дослідження. Зразки крові можуть біти набрані під рутинного набору крові або окремо. Залежно від стану вашого здоров'я, об'єм забраної крові за один раз не буде перевищувати 50 мл. Вам можуть запропонувати здати кров декілька разів. Кількість заборів крові буде

frequency of blood draw will be dependent on your health status but will not exceed 50 mL of blood per week.

Urine: Urine may be collected by discharging this fluid into a container.

Saliva samples: Saliva may be collected by discharging this fluid into a container.

Aspirated Fluids: Aspirated biofluids may be collected as surplus material if you are undergoing a procedure for the purpose of diagnosis or medical treatment. Examples of these procedures include, but are not limited to, ascitic fluid (ascites),aspiration by paracentesis, pleural effusion, cyst, aspiration, synovial fluid and bone marrow aspiration

Use of your samples and data for research purposes:

Your biological samples will be used exclusively for research purposes in accordance with international law in the area of Public Health. They will not be used for transplantation. We have detailed reports from all researchers that receive these samples.

Additional tests and studies to be performed on biological samples may include but not limited to RNA and DNA testing.

Your tissues may be stored in ways that allow the cells to grow and multiply. These multiplying cells may give rise to what is called a cell line. Cell lines can be used for multiple future

залежати від стану вашого здоров'я та не буду перевищувати 50 мл на протязі тижня.

Зразки сечі: зразки сечі збираються спеціальний контейнер.

Зразки слини: зразки слини збираються спеціальний контейнер.

Аспіраційні рідини: можуть збиратися як надлишковий матеріал під час проведення процедури для визначення вашого діагнозу або під час лікування. Аспіраційні рідини включать себе (але не обмежуються): асцитну рідиниу, плевральний випот, кістна рідина , синовіальна рідина або аспірат кісткового мозку.

Використання ваших зразків та даних для дослідницьких цілей:

Ваші біологічні зразки будуть використані виключно із дослідницькою метою згідно міжнародного законодавства в сфері охорони здоров'я. Вони не будуть використані для трансплантації. У нас наявні детальні звіти всіх дослідників, які отримують ці зразки.
Додаткові тести і дослідження, які будуть проводитись на біологічних зразках, можливо, включатимуть тести на РНК і ДНК.

Надані Вами зразки можуть зберігатись в умовах, що дозволять клітинам рости та ділитися. Клітини, що діляться, можуть започаткувати так звані клітинні лінії.. Клітинні лінії в майбутньому можна буде

studies, and these cells may be kept alive for many years.

Patient's signature

_____

Medical information:

For this Study, general data about you and your medical information will be collected. Such information may include: Your diagnosis, treatment, its effectiveness, laboratory tests results and pathology reports. Your personal information (name, surname, address, etc.) will never be given to the researchers and will never be included in any research database.

By signing the enclosed Consent Form, you authorize the placement of your genetic and clinical information in one or more research database(s). This will help to advance medical research by providing researchers with an opportunity to use this information for the purpose of studying diseases, as well as for the purpose of comparing the results obtained by them with the results of other studies. These databases are maintained by medical, academic, government or private institutions. The medical researchers who will have access to your information have professional obligations to keep and maintain confidentiality.

використовувати в різних дослідженнях,, життєдіяльність клітин може зберігатися протягом багатьох років.

Підпис (пацієнт)

_____

Медична інформація:

Для цього дослідження потрібні загальні відомості про Вас і Ваша медична інформація. Вона може включати: Ваш діагноз, лікування і його ефективність, висновки і результати лабораторних і морфологічних досліджень. Ваші особисті дані (ім'я, прізвище) ніколи не будуть передані дослідникам і не будуть входити в жодну наукову базу даних.

Підписуючи надану Вам Форму згоди, Ви даєте згоду на розміщення Вашої генетичної і клінічної інформації в одній і більше базах даних. Це сприятиме передовим медичним дослідженням, даючи можливість іншим дослідникам використовувати цю інформацію для вивчення захворювань, а також для порівняння отриманих ними результатів із результатами інших досліджень. Ці бази даних підтримуються медичними, академічними, державними або приватними установами. Медичні дослідники, які матимуть доступ до Вашої інформації, мають професійні зобов'язання щодо захисту і підтримки конфіденційності.

The data obtained in the course of this Study and/or from your medical records may possibly be reviewed or examined by auditors, medical personnel, and, possibly, government organizations, which may be deciding on the approval of a new method for treatment or diagnostics of the disease. Any review or examination of data is for the purpose of guaranteeing the correctness of the information collected in the course of the Study and the proper conduct of the Study. These examinations will be carried out only in the hospital where the medical records are kept.

Additional information:

By signing this Form, you consent to the collection and use of your samples, as well as access to your medical data for five years from the date of signing this consent. This will help researchers in the further study of your disease. You also may inform your relatives about your decision to take part in this Study.

Benefits from the Study:

By signing this Form, you will not receive any direct benefit from participating in this Study. Neither you, nor your doctors or hospital will receive any other information obtained in the course of this Study, which is conducted with the use of your samples. If in the future some new methods of

Дані, отримані в ході цього Дослідження і/або із Ваших медичних записів, можуть бути надані для розгляду і перевірки аудиторам, медичним співробітникам і, можливо, державним органам, які, можливо, будуть приймати рішення щодо схвалення нового способу лікування захворювання. Будь-який перегляд або перевірка даних має на меті гарантувати точність інформації, зібраної протягом Дослідження, що забезпечує проведення Дослідження належним чином. Такі перевірки проводитимуться тільки в медичному закладі, де зберігаються медичні записи.

Додаткова інформація:

Підписавши цю Форму, Ви надаєте свою письмову згоду на забір і передачу ваших зразків, а також на доступ до Ваших медичних даних протягом 5 років з дати підписання Вами цього документу. Це сприятиме подальшому вивченню Вашого захворювання. Ви також можете повідомити своїх близьких родичів про Вашу згоду прийняти участь у Дослідженні.

Вигоди від участі в дослідженні:

Погодившись надати такі проби, Ви не отримаєте ніяких прямих вигод. Ні Ви, ні Ваші лікарі, ні лікарня не матимуть інформації, отриманої в ході Дослідження, проведеного із використанням Ваших зразків. Однак, якщо в подальшому будуть знайдені нові методи виявлення і

diagnosis and treatment of cancer might be discovered, they may be applied in specialized medical institutions for diagnosis and treatment of such diseases in the regular manner

лікування онкологічних захворювань, вони можуть бути використані спеціалізованими медичними закладами для діагностики і лікування таких захворювань у загальновстановленому порядку.

Risks of the Study:

Ризики, пов'язані із дослідженням:

This Study (including the procedure of taking the samples) does not cause any additional medical risks or pain while samples collection. However, in some cases, taking blood samples may cause bruising or swelling, minor discomfort, infection, vein inflammation or fainting.

Це дослідження (в т.ч. процедура забору зразків) не тягне за собою жодних додаткових медичних ризиків чи додаткових больових відчуттів при заборі зразків. Проте в деяких випадках, забір крові може викликати виникнення синців чи набряку в місці маніпуляції, незначний дискомфорт, інфекцію, запалення вени або запаморочення.

There are no known risks to providing urine. There are no additional medical risks associated with donation of biofluids because you may be undergoing these procedures for the purpose of diagnosis or medical treatment. It is highly unlikely that as a result of the study you will sustain any injuries. Therefore, you will not receive any compensation or medical care through the study. If any such injury should occur, your doctor and the hospital where you are receiving treatment will be responsible for your care.

Дуже малоймовірно, що в результаті Дослідження Ви отримаєте які-небудь ушкодження під час здачі зразків, оскільки ви будете проходити цю процедуру під час діагностики чи вашого лікування. Тому Ви не отримаєте жодної компенсації чи медичної допомоги за рахунок Дослідження. За надання Вам допомоги, у випадку необхідності, відповідальними є Ваш лікар і лікарня, де Вас лікують

Genetic Study Risks:

Ризики генетичних досліджень:

Дане дослідження передбачає проведення генетичних тестів і

This research will involve genetic studies and information. Although, procedures have been put into place that are designed to make it very difficult for the results from genetic research to be linked to you.

отримання генетичної інформації. Проте процедура проведення таких тестів була розроблена таким чином, що зіставити результат тесту із Вашими особовими даними є практично неможливим.

Privacy Risks:

All organizations participating in this study have taken reasonable steps to keep your research data confidential, to the extent permitted by law. Although your genetic information is unique there can be parts of genetic information which are identical for you, your children, parents, siblings, or other relatives. Though it can happen that their genetic information may be used to identify you and vice versa.

Ризик порушення конфіденційності:

Всі організації, залучені до даного дослідження, вжили усіх необхідних заходів для збереження конфіденційності результатів Ваших тестів тією мірою, якою дозволяє закон. Проте абсолютну конфіденційність не можливо гарантувати. Хоча ваша генетична інформація унікальна для вас, ви володієте частинами генетичної інформації, які ідентичні між вами і вашими дітьми, батьками, братами і сестрами, а також іншими родичами. Отже, можливо, що їх генетична інформація може бути використана для визначення вашої особистості і навпаки.

Is my participation mandatory?

Your participation in this Study is voluntary. You may withdraw from participation in the Study at any time, without explaining the reason for it and without any consequences for your further treatment.

Чи повинен я брати участь?

Ваша участь у Дослідженні цілком добровільна. Ви можете відмовитись від участі у Дослідженні на будь-якому етапі без пояснення причини і будь-яких наслідків для Вашого подальшого лікування.

Alternative to participation:

Альтернатива участі:

This Study is conducted solely for research purposes. As an alternative, you may choose not to participate in this Study. Whether you choose to withdraw from this study then any remaining after diagnosing samples will be destroyed.

Costs of the Study:

You do not bear any costs for participation in the Study.

Compensation for the Study:

You will not receive any remuneration for your participation in the Study. If in the future, any research associated with this Study leads to the release of commercial products, neither you, your family, your heirs nor your doctor or hospital will receive any direct benefit, financial or otherwise. The hospital will be given equipment and supplies for its participation in this Study.

Duration of participation in the Study:

Your active participation in this Study will end as soon as all of your samples are obtained. However, your passive participation in the Study will continue for an indefinite period of time, as your biological samples will be stored for an indefinite length of time.

Дане Дослідження проводиться виключно із науковою метою. Альтернативою для Вас є відмова від участі в даному дослідженні. Якщо Ви відмовитесь від участі в даному Дослідженні, тоді всі взяті зразки, що залишилась після проведення діагностики, буде утилізовано.

Витрати на дослідження:

Ви не несете жодних витрат за участь у Дослідженні.

Компенсація за дослідження:

Ви не отримаєте жодної винагороди за участь у Дослідженні. Якщо в подальшому будь-яке Дослідження, пов'язане з даним дослідженням, призведе до випуску комерційної продукції, ні Ви, ні Ваші спадкоємці, ні Ваш лікар, ні лікарня не отримають ніяких прямих вигод. Лікарня отримує обладнання і витратні матеріали для проведення цього Дослідження.

Тривалість участі у дослідженні:

Ваша активна участь у цьому Дослідженні буде припинена відразу після отримання від Вас усіх біологічних зразків. Проте Ваша пасивна участь у дослідженні буде тривати невизначений термін, адже Ваші

Removal from the Study:

You may leave the Study at any time. In such an event, all of your unused biological fluid, tissue or other samples will be destroyed at your request. If the samples have been sent to the researcher, it will not be possible to dispose of them. Your doctor or hospital may discontinue your participation in the Study without your consent.

Who organizes the study?

This study is organized by company "EW Biopharma", acting sui juris and by order of company "Contract Research" If you give your written consent to participate in this Study, you also consent to the transfer of ownership to your samples to the company a daughter company of "Aldima", who may use these samples for further studies without any limitations.

Questions:

If you have any questions concerning this Study, you may contact

_____

at:_____

біологічні зразки зберігатимуться невизначено довгий час.

Вилучення із дослідження:

Ви можете відмовитись від участі у дослідженні на будь-якому етапі. В цьому випадку всі невикористані зразки біологічних рідин, тканина або інші зразки будуть знищені за Вашим запитом. Якщо зразки вже передані досліднику, вилучити їх буде неможливо. Ваш лікар або лікарня можуть припинити Вашу участь у Дослідженні без Вашої згоди.

Хто організує цю роботу?

Цю роботу організовує компанія «ІВ Біофарма», яка діє від свого імені за дорученням компанії «Contract Research». Якщо Ви надасте свою письмову згоду на використання своїх зразків, то Ви погоджуєтесь передати всі права (право власності і майнове право) на такий матеріал компанії «"ІВ Біофарма», яка діє від свого імені за дорученням компанії «Contract Research», яка без жодних обмежень зможе використовувати ці зразки для подальших досліджень.

Питання:

Із будь-яким запитанням щодо цього дослідження Ви можете звернутись до

_____

а телефоном:_____

Patient's signature_____

Підпис
(пацієнт)_____

The patient must fill out the entire sheet on his/her own (Mark what is necessary):

Пацієнт повинен заповнити весь цей листок самостійно (Виділіть необхідне):

Did you read and understand the information provided above? Yes/No

Чи прочитали і зрозуміли Ви Інформацію для пацієнтів? Так/Ні

Did you have an opportunity to ask questions and discuss the reason why your blood, tissue or other samples are requested? Yes/No

Чи мали Ви можливість поставити питання і обговорити причину, з якої виникла потреба у пробах Вашої крові, тканини або інших зразків? Так/Ні

Did you get full answers to all of your questions? Yes/No

Чи одержали Ви вичерпні відповіді на всі питання? Так/Ні

Did you receive sufficient information about the reasons for requesting these samples? Yes/No

Чи одержали Ви достатньо інформації про причини необхідності у цих зразках? Так/Ні

Do you understand that you may choose not to provide your blood, tissue or other samples without giving any reasons for it and without consequences for your further medical care? Yes/No

Чи зрозуміло Вам, що Ви можете відмовитись від надання Вашої крові, тканини чи інших зразків без пояснення причини і будь-яких наслідків для Вашого подальшого лікування ? Так/Ні

Do you agree to the use of your samples and medical data for research and development as described above? Yes/No

Чи даєте Ви згоду на використання Ваших проб та клінічної інформації для досліджень і розробок, як це було описано вище? Так/Ні

For future studies, if possible, by decision of the researcher? Yes/No

для подальших досліджень, якщо це буде можливо, згідно з рішенням дослідника? Так/Ні

Did you understand and do you consent to the clinical data about you being maintained on a confidential basis? However, the appropriate clinical data may be provided, on a confidential basis, to supervisory authorities and commercial companies interested in any Study results or in any tests and methods of treatment? Yes/No

Чи зрозуміли Ви і чи згодні з тим, що клінічні відомості стосовно Вас матимуть конфіденційний характер? Проте відповідні клінічні дані можуть бути надані на конфіденційній основі наглядовим органам і комерційним компаніям, зацікавленим у будь-яких результатах досліджень або в будь-яких тестах і методах лікування?Так/Ні

Do you understand and do you consent to the Study results obtained with the use of your biological samples being possibly published, although in such an event you will not be identified individually? Yes/No

Чи зрозуміли Ви і чи згодні з тим, що отримані результати досліджень з використанням Ваших зразків можуть бути опубліковані, хоч при цьому Ви не будете вказані як індивідуальна особа? Так/Ні

Do you give your consent to having the researchers in the future submit a sample collected from you to other researchers studying various diseases, including cancer? Yes/No

Чи дасте Ви свою згоду на те, щоб в подальшому дослідники передавали проби, взяті у Вас, іншим дослідникам, які вивчають різні захворювання, включаючи рак? Так/Ні

Do you understand that you transfer all rights, title and vested interest to your blood, tissue or other types of sample to the company "EW Biopharma", acting sui juris and by order of company "Contract Research" ? Yes/No

Чи зрозуміли Ви, що Ви передаєте всі права, право власності і майнове право на пробу Вашої крові компанії «ІВ Біофарма», яка діє від свого імені за дорученням компанії «Contract Research»? Так/Ні

Indicate the types of samples you agree to donate in the table below:

Позначте в таблиці типи зразків, які Ви згодні надати:

Surplus tissue (surgically removed)	Yes	No

Надлишкова тканина (видалена хірургічно)	так	ні

Blood Sample	Yes	No	Зразки крові	так	ні
Salvia	Yes	No	Рідина зразки сечі	так	ні
Ascitic Fluid	Yes	No	Слина	так	ні
Pleural effusion (surplus)	Yes	No	Асцитні рідини (надлишкові)	так	ні
Bone Marrow Aspiration (surplus)	Yes	No	Плевральний випіт	так	ні
Cyst Aspiration (surplus)	Yes	No	Кістковий мозок (надлишковий біопсій ний матеріал)	так	ні
Synovial Fluid (surplus)	Yes	No	кістний аспірат	так	ні
			Синовіальна рідина	так	Ні

Do you agree with mentioned above? Yes/No

Чи згодні Ви із вищевказаним? Так/Ні

Full name of patient:

ПІБ пацієнта:

Patient ID:

ID пацієнта:

My signature below means that I have read or I have been read, in a language understandable to me, all of the information contained in this consent form. The meaning of this information has been explained and is completely clear to me. I have had the time and opportunity to ask questions about the Study. I have received answers to all of my questions. I have read all pages of this consent form, including the risk and compensation sections. I certify that I am eighteen (18) years of

Мій підпис нижче означає, що я прочитав або мені прочитали зрозумілою мені мовою всю інформацію, яка міститься в цієї Формі Згоди. Значення цієї інформації мені було роз'яснено і цілком зрозуміле. Я мав(ла) час і можливість поставити питання про дослідження. На всі питання я отримав(ла) відповідь. Я прочитав(ла) всі сторінки цієї форми згоди, у тому числі розділи про ризик і винагороду. Я засвідчую, що мені виповнилось вісімнадцять (18) або більше років. Я також засвідчую, що

age or older. I also certify that all of the information that I have provided, including my medical data, is true and accurate to the best of my knowledge. By signing this form, I do not waive any of my legal rights. I will receive a signed copy of this form.

вся інформація, яку я надав, у тому числі мої медичні дані, наскільки мені відомо, вірні і точні. Я не відмовляюсь від жодного із моїх законних прав, підписуючи цю форму. Я отримаю підписану копію цієї форми

Signature (patient)

_____

_____

Підпис (пацієнт)

_____

_____

Date: _____

Time: __ : ___

Дата: _____

Час : ___ : ____

I have explained the essence of the study to the above patient and he/she has expressed willingness to participate

Я пояснив суть Дослідження вказанному вище пацієнтові і він\вона виявив\ла бажання взяти участь.

Doctor's signature:

_____

Date: _____

Time:__ : __

Підпис лікаря

_____

Дата:_____

Час: ____:____

Full name

_____

ПІБ лікаря:

_____

Witness:

_____

Свідок:

_____

Date: _____

Time: ____ : ___

Дата: _____

Час: __ : __

ACCEPTED AND APPROVED

THE ORGANIZER –

ПРИЙНЯТО ТА ПОГОДЖЕНО:

ОРГАНІЗАТОР –

Limited Liability Company	Товариство з обмеженою відповідальністю ТОВ
Ukraine Association of Biobank	Українська Асоціація Біобанків в особі Голови:
Head UAB Alekseenko Mykola	Алексеенко Миколи
Signature	Підпис
Date	Дата

Annex 3: Standard Operating Procedure Collection of Plasma from Whole Blood

	Standard Operating Procedure Collection of Plasma from Whole Blood
SOP – 03 Version 1.2	Effective Date: 25.05.2016

1.0 *Purpose*
   This document describes the process for the collection of human plasma from whole blood.

2.0 *Scope*
   These guidelines apply to personnel intending to preserve plasma biobanking studies

3.0 *Requirements*
   3.1 <u>Equipment</u>
       3.1.1 Centrifuge with swinging bucket rotor
       3.1.2 −80 °C freezer
       3.1.3 Biosafety cabinet Hood
       3.1.4 Pipette aid

3.2 <u>Materials and Equipment</u>

3.2.1 A refrigerated (2–8 °C) centrifuge capable of spinning at

Fig. A.1 K_2 EDTA BD Vacutainer®

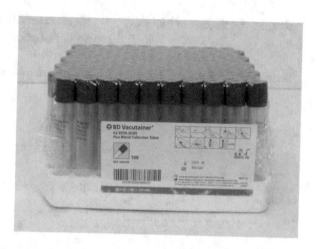

Fig. A.2 Freezing tube storage

1500–2000 g that will hold 16 × 100 mm tubes

3.2.2 10 mL lavender-top K_2 EDTA BD Vacutainer® venous blood collection tubes and 10 mL red-top serum BD Vacutainer® venous blood collection tubes (Fig. A.1)

Disposable transfer pipettes

VWR Low Temperature 5 mL Freezer vials, Sterile (Product # 16001-104). Serum and plasma from the BD collection tubes are transferred to these secondary tubes for freezing and storage (Fig. A.2).

3.2.3 Freezer and cryobox for storing freezer vials. Ultra-cold freezers at or below −70 °C are preferred.

4.0 *Method*

Pre-chill 10 mL lavender-top K_2 EDTA BD Vacutainer® venous blood collection tubes on wet ice for at least 5 min.

Follow standard phlebotomy practices and procedures, including wearing gloves during venipuncture and when handling blood collection tube and secondary tube to minimize exposure hazard. Consult product inserts for details of collection procedures, and order of collection when also collecting samples for other purposes.

Immediately after allowing the lavender-top Vacutainer® tube to completely fill, gently invert the tube 8 times, and return the tube to the wet ice bath

Within 30 min of collection, centrifuge at 1500–2000 g for 10 min in a refrigerated centrifuge (2–8 °C).

Within 30 min of centrifugation carefully transfer the plasma into a 5 mL VWR freezer vial. Do not attempt to retrieve the last 0.2 mL of plasma.

Plasma samples should be immediately frozen at −70 to −80 °C, and shipped on dry ice to a designated facility.

Deviations from the prescribed collection protocol should be reported to the designated coordinator.

Annex 4: Ethics Committee Review Form

<table>
<tr>
<td>

The Ukraine Association of Biobank
ETHICS COMMITTEE REVIEW FORM
The EC is organized and operates in compliance with the ICH GCP
requirements and Ukraine Law

</td>
<td>

Application No: 1506-17
Date Received:
15/06/17

</td>
</tr>
</table>

1. TITLE OF PROJECT

> Blood and Blood components, Frozen tissues and FFPE tissue blocks collection _for further
> diagnostic studies

2. APPROVAL DETAILS
   What is the Ethical Review Number (ERN) for the project

   Protocol/Study No: 1506/2017

3. THIS PROJECT IS:

   The Ukraine Association of Biobanks

4. INVESTIGATORS

   a) PLEASE GIVE DETAILS OF THE PRINCIPAL INVESTIGATORS OR SUPERVISORS (FOR PGR
      UAB PROJECTS)

Name: Title / first name / family name	Attikov Volodimir
Highest qualification & position held:	Medical Director
School/Department	Ukraine Association of Biobank
Telephone:	+38095-35-64-777
Email address:	

Name: Title / first name / family name	Alyeksyeyenko Mykola
Highest qualification & position held:	President
School/Department	Ukraine Association of Biobank
Telephone:	+380504007080
Email address:	n.alecseenko@gmail.com

5. ETHIC COMMITTEE MEMBERS ATTENDED:

   Proff. Mitriaeva N.A., Proff. Popovskai L.D., PhD. Popova L.D., PhD. Pasiahsvili L.V., PhD. Sherban N.G.,
   PhD. Gukov V.I., PhD. Starenkii V.P., PhD. Vinnikov V.A., PhD. Gramatiuk S.N., MD. Korzov A.E., Dr.
   Attikov V., Medical lawyer Alyeksyeyenko M.I.

6. DEVELOPED:
   Ukraine Association of Biobank, Kharkiv 61018, Balakireva str. 1, T. +38050-400-70-80, email:
   n.alecseenko@gmail.com

7. Following documents have been submitted to the Ethics Committee on date: 15/06/2017

8. Low Risk application information:
   Research may be considered low risk when it arises from
   a. Masters or PhD theses where the projects do not raise any issue of deception, threat,
      invasion of privacy, mental, physical or cultural risk or stress, and do not involve gathering
      personal information of a sensitive nature about or from individuals.

b. Masters or PhD level supervised projects undertaken as part of specific course requirements where the projects do not raise any issue of deception, threat, invasion of privacy, mental, physical or cultural risk or stress, and do not involve gathering personal information of sensitive nature about or from individuals.

c. Undergraduate and Honours class research projects which do not raise any issue of deception, threat, invasion of privacy, mental, physical or cultural risk or stress, and do not involve gathering personal information of sensitive nature about or from individuals, but do not have blanket approval as specified in Section 4 of the Principles and Guidelines.

9. **REVIEWED** YES NO

Protocol version	1506/2017-1
Patient Consent Form version	1506/2017-1CF
Investigator's Brochure version	1506/2017-1BV
Amendment(s) number	
Insurance number	
Regulatory approval protocol	
CV of Investigator(s)	

10. **The study was:**
Approved Conditionally approved (see attached letter)
Deterred pending further information (see attached letter)
Rejected (see attached letter)
Is review of study status required Yes No If Yes._____
Annually Periodically (when) _____
The investigator(s) is/are EC member(s) and therefore did not vote none

Date: 15/06/2017

Chairman of EC _____ Proff. Mitriaeva N.A. _____ Signature

Secretary _____ MD. Korzov Aleksey _____ Signature

Medical lawyer _____ Alyeksyeyenko Mykola _____ Signature

Printed in the United States
by Baker & Taylor Publisher Services